ANNALES ACADEMIÆ SCIENTIARUM FENNICÆ

MATHEMATICA

DISSERTATIONES

151

UNIFORMLY QUASIREGULAR MAPPINGS ON ELLIPTIC RIEMANNIAN MANIFOLDS

RIIKKA KANGASLAMPI

Helsinki University of Technology, Department of Mathematics and Systems Analysis

Dissertation for the degree of Doctor of Science in Technology to be presented, with due permission of the Faculty of Information and Natural Sciences, for public examination and debate in Auditorium K at Helsinki University of Technology (Espoo, Finland) on the 30th of May, 2008, at 12 o'clock noon.

HELSINKI 2008
SUOMALAINEN TIEDEAKATEMIA

ISSN 1239-6303
ISBN 951-41-0989-9

Received 29 April 2008

2000 Mathematics Subject Classification:
Primary 30C65; Secondary 53C20, 57M12, 30D05.

YLIOPISTOPAINO
HELSINKI 2008

Abstract

In this thesis we study uniformly quasiregular (abbreviated uqr) mappings on compact riemannian manifolds. We prove that the Julia set $\mathcal{J}_f$ of a uqr-mapping $f : M^n \to M^n$ on a compact riemannian manifold M^n is non-empty. We extend the rescaling principle from euclidean spaces to families of quasiregular mappings between a euclidean space and a riemannian manifold. Thus we can use the rescaling principle to obtain from the family (f^j) of the iterates of the uqr-mapping f on the manifold M a quasiregular mapping $g : \mathbb{R}^n \to M^n$ defined in the whole space $\mathbb{R}^n$. Combining these results, we notice that if there exists a uqr-mapping on a compact riemannian manifold M^n, there exists a quasiregular mapping $g : \mathbb{R}^n \to M^n$. In other words, the manifold M^n is quasiregularly elliptic. The converse result is proved in three dimensions: we construct a uqr-mapping on each oriented quasiregularly elliptic 3-dimensional compact riemannian manifold.

Tiivistelmä

Tässä tutkimuksessa tarkastellaan tasaisesti kvasisäännöllisiä (lyh. tks) kuvauksia kompakteilla Riemannin monistoilla. Työssä todistetaan, että kompaktin Riemannin moniston M^n tks-itsekuvauksen $f : M^n \to M^n$ Julian joukko $\mathcal{J}_f$ on epätyhjä. Seuraavaksi laajennetaan skaalausperiaate euklidisen avaruuden ja Riemannin moniston välisten kvasisäännöllisten kuvausten perheille. Skaalauksen avulla saadaan moniston M tks-itsekuvauksen f iteraattien jonosta (f^j) muodostettua koko avaruudessa $\mathbb{R}^n$ määritelty kvasisäännöllinen kuvaus $g : \mathbb{R}^n \to M^n$. Näin ollen havaitaan, että tks-itsekuvauksen olemassaolosta kompaktilla Riemannin monistolla M seuraa, että on olemassa kvasisäännöllinen kuvaus $g : \mathbb{R}^n \to M^n$. Tämä tarkoittaa, että monisto M^n on kvasisäännöllisesti elliptinen. Kolmiulotteisille kompakteille Riemannin monistoille sama väite todistetaan myös käänteiseen suuntaan: jokaiselle suunnistuvalle kvasisäännöllisesti elliptiselle kompaktille kolmiulotteiselle Riemannin monistolle konstruoidaan tks-itsekuvaus.

Acknowledgements

This research has been carried out at the Institute of Mathematics at Helsinki University of Technology during the years 2003-2008. The work has been partly supported by the Finnish Cultural Foundation and the Graduate School in Mathematical Analysis and Its Applications.

I would like to thank my advisor, docent Kirsi Peltonen, for introducing me to uniformly quasiregular mappings and for being a rich source of ideas throughout these years. I am also grateful to my supervisor, professor Olavi Nevanlinna, for encouraging me to study uniformly quasiregular mappings, even though that is not one of the main fields of study in our department. Professor Aimo Hinkkanen at the University of Illinois at Urbana-Champaign and PhD Pekka Pankka at the University of Michigan have given me valuable comments about this work. Thanks to their efforts, this monograph is much better than it would have otherwise been.

My dear colleagues, friends and family: I wish to thank you all for believing that I can do this, and for telling it to me again and again. Without you, I would have given up a long time ago.

Espoo, April 2008

Riikka Kangaslampi

Contents

1. Introduction

1.1. History of the research on quasiregular mappings. The first attempts to extend the theory of analytic functions from the complex plane were made already in the 1920's. In 1928, H. Grötzsch considered homeomorphisms which do not preserve angles or circles but do not modify them drastically, either. The theory of quasiconformal mappings was born. This theory was developed further by L.V. Ahlfors, O. Teichmüller, O. Lehto and K.I. Virtanen, and, with some new results by M. Lavrent'ev, it was generalized to the space $\mathbb{R}^n$, $n \geq 2$, in 1938.

The theory of non-bijective analytic functions could be generalized as well, and the so-called quasiregular mappings were obtained. Yu. G. Reshetnyak started the systematic study of these mappings in his series of articles published between 1966 and 1969. He proved basic results for quasiregular mappings using the theory of non-linear partial differential equations, Sobolev spaces and differential geometry. In the articles [MRV1], [MRV2] and [MRV3] published in 1969-1972, O. Martio, S. Rickman and J. Väisälä continued the study of quasiregular mappings using also other methods, such as the theory of modulus of curve families.

In the last 35 years, the theory of quasiregular mappings has expanded to a diverse field of research. Quasiregular mappings have turned out to be a natural way of generalizing some aspects of the theory of complex functions into higher dimensions. Even though not all the characteristics of analytic functions hold for quasiregular mappings, there are enough profound results, such as the theorems of Picard and Montel [R2].

A new, surprising class of mappings was introduced by T. Iwaniec and G. Martin in 1996: uniformly quasiregular mappings [IM2]. These are quasiregular mappings with a common distortion bound for all the iterates. The dynamics of uniformly quasiregular mappings correspond to the dynamics of rational mappings on the plane. In dimension 2 they are just rational mappings (up to a quasiconformal change of variables). We can define the Fatou and Julia sets of uniformly quasiregular mappings and use the same terminology as for Möbius mappings to classify their fixed points. Examples of uniformly quasiregular mappings in higher dimensions are mappings of the Lattès type in $\mathbb{S}^3$ [Ma2] and, more generally, mappings constructed by the trap method on spherical manifolds [Pe].

This study of uniformly quasiregular mappings is motivated by the open question of which manifolds are quasiregularly elliptic. By a quasiregularly elliptic manifold we mean such an oriented riemannian manifold M that there exists a non-constant quasiregular mapping $f : \mathbb{R}^n \to M$. Elliptic manifolds were originally introduced by M. Gromov [Gr], but with some new results they have reappeared in discussion in recent years. In 2001, M. Bonk and J. Heinonen found a cohomology obstruction for ellipticity [BH]. In addition hyperbolic manifolds are not elliptic, since their fundamental groups grow exponentially, and the growth for elliptic manifolds can be at most polynomial [VSC, Theorem X.5.1]. In dimension 3, J. Jormakka has characterized all elliptic manifolds, but much less is known about higher dimensions. One interesting example is the connected sum $\mathbb{S}^2 \times \mathbb{S}^2 \# \mathbb{S}^2 \times \mathbb{S}^2$, which S. Rickman recently showed to be quasiregularly elliptic [R3]. It is not known whether this manifold supports a uniformly quasiregular mapping or not.

1.2. **Overview of this thesis.** This thesis consists of three parts essentially. In the first two chapters we give some preliminary results about coordinate charts on riemannian manifolds, define uniformly quasiregular mappings and show some of their properties. The following two chapters consider manifolds which support a uniformly quasiregular mapping. We prove that the Julia set is non-empty and these manifolds are quasiregularly elliptic. In the last two chapters a converse result is proved to hold in three dimensions: all quasiregularly elliptic 3-manifolds support a uniformly quasiregular mapping.

In chapter 2 we prove that we can always choose a bilipschitz-continuous coordinate chart for a riemannian manifold. This result is commonly known, but an elementary proof could not be easily found in the literature. After these preliminary results about coordinate charts, we define quasiregular mappings and uniformly quasiregular mappings as their subset in chapter 3. Quasiregular mappings are first defined on euclidean spaces and then more generally on n-dimensional riemannian manifolds. We also define branch sets and normal families and give some of their properties.

We define Julia sets in chapter 4, and prove that the Julia set of a uniformly quasiregular mapping is non-empty. We need this result to generalize the rescaling principle to mappings from the euclidean space to riemannian manifolds in chapter 5. The fact that the Julia set is non-empty was stated, for example, by Iwaniec and Martin in [IM1] for mappings on euclidean spaces, but a proof for closed riemannian manifolds has not been presented earlier. With the help of the generalized version of the rescaling lemma (also known as Zalcman's lemma and Bloch-Brody principle in the literature), we are able to show that if a riemannian manifold supports a uniformly quasiregular mapping, this manifold is quasiregularly elliptic.

The last two chapters are devoted to 3-dimensional manifolds. In chapter 6 we explain how some 3-manifolds can be represented by polygons and what we mean by model geometries. Then in chapter 7 we construct a uniformly quasiregular mapping on each elliptic 3-manifold. We need to consider the so-called spherical space forms, euclidean space forms and manifolds covered by $\mathbb{S}^2 \times \mathbb{R}$ separately. As a result of this thesis, we have thus shown that in all dimensions, compact riemannian manifolds supporting a uniformly quasiregular mapping are quasiregularly elliptic and that in 3 dimensions, compact elliptic riemannian manifolds support a uniformly quasiregular mapping.

2. PRELIMINARY RESULTS

In the study of riemannian manifolds, and especially when trying to generalize results from euclidean spaces to riemannian manifolds, coordinate charts play an important role. The better-behaved they are the better. For our purposes, bilipschitz continuity is enough. It allows us to define quasiregular mappings on riemannian manifolds and is of great importance later in the proofs. Therefore, in this chapter we define geodetically convex neighbourhoods, prove that such a neighbourhood exists at each point of a riemannian manifold, and show that we can choose such an atlas for our manifold that all the coordinate charts are locally bilipschitz-continuous. From now on we always consider smooth and connected riemannian manifolds without boundary.

2.1. **Normal coordinates.** Let us define normal coordinates as in [Lee]. Let M be a riemannian manifold and $p \in M$ a point. Then the restricted exponential map $\exp_p$ maps an open subset V_p of the tangent space T_pM into the manifold M.

Lemma 2.1 (Normal neighbourhood lemma). *For any $p \in M$, there exists a neighbourhood V of the origin in T_pM and a neighbourhood U of p in M such that $\exp_p : V \to U$ is a diffeomorphism.*

A detailed proof for this lemma can be found in [Lee, p. 76]. Let us define normal neighbourhoods as follows:

Definition 2.2. Any open neighbourhood U of a point $p \in M$ is called a *normal neighbourhood*, if it is the diffeomorphic image under $\exp_p$ of a star-shaped open neighbourhood of $0 \in T_pM$.

Let $\{E_i\}$ be an orthonormal basis of the tangent space T_pM. Then the mapping

$$E(x^1, \ldots, x^n) = x^i E_i$$

is an isomorphism $E : \mathbb{R}^n \to T_pM$. If U is a normal neighbourhood of $p \in M$, we can combine this isomorphism with the exponential map to get a coordinate chart

$$\varphi := E^{-1} \circ \exp_p^{-1} : U \to \mathbb{R}^n.$$

Any such coordinates are called *Riemannian normal coordinates*, centred at the point p. Given $p \in M$ and a normal neighbourhood U of p, there is a one-to-one correspondence between normal coordinate charts and orthonormal bases at p.

The following useful properties hold for normal coordinates [Lee, p. 78]:

Theorem 2.3. *Let $(U, (x^i))$ be any normal coordinate chart centred at p. Then*

(1) for any $V = V^i \partial_i \in T_pM$, the geodesic γ_V starting at p with initial velocity vector V is represented in normal coordinates by the radial line segment

$$\gamma_V(t) = (tV^1, \ldots, tV^n),$$

as long as γ_V stays within U.

(2) the coordinates of p are $(0, \ldots, 0)$.

(3) *the components of the metric at* p *are* $g_{ij} = \delta_{ij}$.
(4) *the first partial derivatives of* g_{ij} *and Christoffel symbols* $\Gamma^i_{jk}(p)$ *vanish at* p.

2.2. **Geodetically convex neighbourhood.** Distance μ_U induced by the restriction of the riemannian metric of M to set $U \subset M$ does not necessary agree with the original distance μ. If we wish to make sure that the distances agree, the neighbourhood U has to be geodetically convex. Let us now consider the requirements for this to happen.

Definition 2.4. A subset V of a riemannian manifold M is *geodetically convex* if for any p, $q \in V$ there exists a minimal segment between p and q which stays in V. By a minimal segment we mean a segment of a geodesic, which minimizes the arc distance between its two endpoints.

Definition 2.5. An open ball $\mathcal{B}(p, r_0)$ (radius r_0, centre at $p \in M$) is *locally geodetically convex* if every sphere $\mathbb{S}(p, r)$ centred at p with radius $r < r_0$ satisfies the following condition: if γ is a geodesic tangential to the sphere $\mathbb{S}(p, r)$ at $\gamma(0)$, then

$$\mu(p, \gamma(u)) \geq r$$

for any u small enough.

The following theorem was originally presented by J. H. C. Whitehead. It is proved, for example, in [Hi], and we will here follow the idea of that proof.

Theorem 2.6. *Let* M *be a riemannian manifold and* D *a smooth connection on* M. *Then for any point* $p \in M$ *there exists a neighbourhood* U *of* p *which is geodetically convex.*

We will prove Theorem 2.6 by three lemmas. The last of these lemmas shows the existence of the desired neighbourhood.

We can assume that the torsion of the connection D is zero, since there exists a unique torsion-free connection with the same geodesics as the original connection [Hi, p. 65]. Let us also point out that since the claim is local, we can restrict our considerations into a coordinate neighbourhood of one arbitrary point $m \in M$.

Let us choose a normal coordinate representation $x^1, x^2, \ldots, x^n$ for the point m in a neighbourhood $A \subset M$. Thus $x^i(m) = 0$ and the Christoffel symbol $\Gamma^i_{jk}(m) = 0$ for any i, j and k (see [Lee, p. 78] for these and other properties of normal coordinates). Let us define a local metric $d(p, q)$ in the neighbourhood A as follows:

$$d(p, q) := \big(\sum_i (x^i(p) - x^i(q))^2\big)^{1/2}.$$

Let us also define $f(p) := d(p, m)$ for $p \in A$. This function gives the euclidean distance from the point m. We wish to consider only the interior of the set A, so let from now on $B(p, c) := \{q \in A \ : \ d_A(p, q) < c\}$ for any $p \in A$. Thus $B(p, c)$ is the restriction to the set A of an ordinary ball of radius c centred at p.

Let $p \in A$ and let $\hat{N}$ be such an open subset of the tangent bundle TM that the exponential mapping $\exp_p$ is defined in $\hat{N}$. There exists a radius $r_p > 0$ such that the mapping $\mathcal{G} : \hat{N} \to M \times M$, $\mathcal{G}(p, Y) = (p, \exp_p Y)$, is a diffeomorphism on the set $\{(q, X) : q \in B(p, r_p), d(q, \exp_p X) < r_p\}$ [Hi, Corollary 2, chapter 9.3]. Let $c > 0$ be such that $\bar{B} = \bar{B}(m, c) \subset A$ and $\bar{B}$ is compact. Then for any $p \in \bar{B}$ there exists a radius r_p with the properties presented above. The family $\{\bar{B}(p, r_p) : p \in \bar{B}\}$ of the neighbourhoods is a cover for the compact set $\bar{B}$. Therefore, a finite subcover consisting of neighbourhoods of some points $p_1, \ldots, p_k$ can be chosen.

Define $s = \min\{r_1, \ldots, r_k\}$, where $r_j = r_{p_j}$. Then, for any $p \in \bar{B}$ the mapping $\exp_p$ takes $\tilde{U}_p$, a ball of radius r in the tangent space T_pM, centred at the origin, diffeomorphically to the set $B(p, s)$, since any $p \in \bar{B}$ belongs to the set $B(p_j, r_j)$ with some index j, and, therefore, $\mathcal{G}$ is a diffeomorphism in the set $\{(q, X) : q \in B(p_j, r_j), d(q, \exp_p X) < r_j\}$. Set $q = p$. Then, $\exp_p$ is a diffeomorphism from a neighbourhood $\tilde{V}_p$ of the origin in T_pM to the set $B(p, r_j)$, and $s \leq r_j$. We can now define $\tilde{U}_p$ as the preimage of $B(p, s)$ under $\exp_p$.

We have proved the following lemma:

Lemma 2.7. *For any such $c > 0$ that $\bar{B}(m, c) \subset A$, there exists $s > 0$ such that when $p \in \bar{B}(m, c)$, the mapping $\exp_p$ is a diffeomorphism from a ball $\tilde{U}_p$ of radius r, centred at the origin of the tangent space T_pM, onto the set $B(p, s) \subset A$.*

The remaining two lemmas finalize the proof of Theorem 2.6. Remember that we earlier defined the mapping f to be the euclidean distance from the point m, that is, $f(p) = d(p, m)$.

Lemma 2.8. *There exists a real number a, $0 < a < 1$, $\bar{B}(m, a) \subset A$, such that if $0 < b < a$, γ is a geodesic with tangent $T_{\gamma(t)}$, $f \circ \gamma(0) = b$, and $T_{\gamma(0)} f = 0$, then the mapping $f \circ \gamma$ obtains its real local minimum at the point $\gamma(0)$. In other words, near the point $\gamma(0)$ the geodesic γ lies outside the ball $B(m, b)$ if at the point $\gamma(0)$ it is tangential to a ball of radius b, centred at m.*

Proof. Assume that $|T_{\gamma(0)}| = 1$. Denote $T = \dot{\gamma}^i \frac{\partial}{\partial x_i}$ and assume that T is extended to a smooth field near the point $\gamma(0)$.

Since $f = \left(\sum_i (x^i)^2\right)^{\frac{1}{2}}$, it holds that

$$\frac{\partial f}{\partial x^k} = \frac{1}{2}\Big(\sum_i (x^i)^2\Big)^{-\frac{1}{2}} 2x^k = \frac{x^k}{f}.$$

Thus we get

$$Tf = df(T) = \dot{\gamma}^i \frac{\partial f}{\partial x^i} = \sum_{i=1}^{n} \frac{\dot{\gamma}^i x^i}{f}$$

and furthermore,

$$\begin{aligned} T^2 f &= d(Tf)(T) = \dot\gamma^k \frac{\partial(Tf)}{\partial x^k} = \sum_k \dot\gamma^k \Big(\frac{\partial}{\partial x^k} \big(\sum_i \frac{\dot\gamma^i x^i}{f} \big) \Big) \\ &= \sum_k \dot\gamma^k \Big(\frac{\dot\gamma^k}{f} + \frac{1}{f} \sum_i \frac{\partial \dot\gamma^i}{\partial x^k} x^i + \big(\sum_i \dot\gamma^i x^i \big) \frac{-1}{f^2} \frac{x^k}{f} \Big) \\ &= \sum_k \Big(\frac{(\dot\gamma^k)^2}{f} - \frac{\dot\gamma^k x^k}{f^3} \sum_i \dot\gamma^i x^i + \frac{\dot\gamma^k}{f} \sum_i \frac{\partial \dot\gamma^i}{\partial x^k} x^i \Big). \end{aligned}$$

At the point $\gamma(0)$, when $t = 0$, we get from the assumption that $Tf = 0$ and $f(\gamma(t)) = b$. Therefore, for the value $t = 0$, we have

$$T^2 f = \frac{1}{b} \Big(\sum_k (\dot\gamma^k)^2 + \sum_i \dot\gamma^k \frac{\partial\, \dot\gamma^i}{\partial x^k} x^i \Big). \tag{1}$$

Since γ is a geodesic, we have for any i [Hi, p. 58]

$$\ddot\gamma^i + 2G^i(\dot\gamma) = 0, \tag{2}$$

where

$$G^i(\dot\gamma) := \frac{1}{2} \Gamma^i_{jk}(\gamma) \dot\gamma^j \dot\gamma^k.$$

When we apply this definition to the equation (2) and write $\ddot\gamma$ in the form

$$\ddot\gamma^i = \frac{\mathrm{d}}{\mathrm{d}t} \dot\gamma^i(\gamma(t)) = \dot\gamma^k \frac{\partial \dot\gamma^i}{\partial x^k},$$

we get

$$\dot\gamma^k \frac{\partial \dot\gamma^i}{\partial x^k} + \Gamma^i_{jk}(\gamma) \dot\gamma^j \dot\gamma^k = 0. \tag{3}$$

By applying this further to the equation (1) and using the fact $|T_{\gamma(0)}| = 1$, we see that

$$T^2 f = \frac{1}{b} \Big(1 - \sum_i \Gamma^i_{jk}(\gamma) \dot\gamma^j \dot\gamma^k x^i \Big) \tag{4}$$

when $t = 0$. Let us now consider the sum $\sum_i \Gamma^i_{jk}(\gamma) \dot\gamma^j \dot\gamma^k x^i$ more closely. Choose $0 < a < 1$ in such a way that for any point p with $f(p) = d(m,p) \leq a$, the equation $|\Gamma^i_{jk}(p)| \leq \frac{1}{2n^3}$ is true for any indices i, j and k. It is possible to choose such a value a, since Γ^i_{jk} is continuous and $\Gamma^i_{jk}(m) = 0$ for any i, j and k (Theorem 2.3).

Therefore, at the point $\gamma(0)$,

$$\begin{aligned} |\sum_i \Gamma^i_{jk}\dot{\gamma}^j\dot{\gamma}^k x^i| &\leq \sum_i |\Gamma^i_{jk}||\dot{\gamma}^j\dot{\gamma}^k x^i| \\ &\leq \frac{1}{2n^3}(\sum_{i,j,k} 1) \\ &\leq \frac{1}{2}, \end{aligned}$$

where $n = \dim(M)$ and $|\dot{\gamma}^j\dot{\gamma}^k x^i| \leq 1$, since $|x| = |\gamma(0)| = b < a < 1$ and we made the assumption that at the point $\gamma(0)$ the sum $\sum_k \dot{\gamma}_k^2 = |T|^2 = 1$, and so $|\dot{\gamma}^i| \leq 1$ for any i. Thus $T^2 f(\gamma(0)) > 0$, and the function $f \circ \gamma$ obtains a real local minimum at $t = 0$. □

Now we can tackle the last lemma to prove Theorem 2.6:

Lemma 2.9. *Let a be given as in Lemma 2.8. By defining $c = \frac{a}{2}$, we get $s > 0$ from Lemma 2.7. If $s < \frac{2}{3}a$, the set $B(m, \frac{s}{2})$ is geodetically convex.*

Proof. Choose arbitrary points $p, q \in B(m, \frac{s}{2})$. According to Lemma 2.7, the exponential map $\exp_p$ is a diffeomorphism between U_p, a neighbourhood of the origin of the tangent space T_pM, and the set $B(p, s)$. Since $d(p, q) \leq s$, also $q \in B(p, s)$. Thus the exponential map gives a geodesic γ, which is defined on the interval $[0, u]$ and for which $\gamma(0) = p$, $\gamma(u) = q$, and $\gamma(t) \in B(p, s)$ for any $t \in [0, u]$.

We will now prove that this geodesic connecting the points p and q stays inside the ball $B(m, \frac{s}{2})$. It means that $f \circ \gamma(t) < \frac{s}{2}$ for any $t \in [0, u]$. Here $f(l) = d(l, m)$ as before. We prove this by showing that the maximal value of $f \circ \gamma$ is less than $\frac{s}{2}$.

Let $v \in [0, u]$ be the point where $f \circ \gamma$ obtains its maximal value. We know that $f \circ \gamma(v) < a$, since

$$f \circ \gamma(v) = d(m, \gamma(v)) \leq d(m, p) + d(p, \gamma(v)) < \frac{s}{2} + s < a.$$

Assume that $f \circ \gamma(v) \geq \frac{s}{2}$. Since v is a point where a maximal value is obtained, $(f \circ \gamma)'(v) = 0$. We also know that $f \circ \gamma(v) < a$, and, therefore, according to Lemma 2.8, the mapping $f \circ \gamma$ has a real local minimum at v. This is a contradiction to the fact that v is a maximum point. Thus our assumption is false, and $f \circ \gamma(t) < \frac{s}{2}$ for any $t \in [0, u]$. The geodesic connecting the arbitrarily chosen points p and q stays inside $B(m, \frac{s}{2})$, and so the set $B(m, \frac{s}{2})$ is by definition geodetically convex. □

Note that since riemannian manifolds are smooth and a riemannian metric gives a unique smooth torsion-free connection (the Levi–Civita conection) on the manifold [Hi, p. 71], we have in fact proved the following corollary:

Corollary 2.10. *Each point of a riemannian manifold has a geodetically convex neighbourhood.*

2.3. **Locally bilipschitz-continuous coordinate charts.** Let (M,g) be a riemannian manifold and $p \in M$ a point. Choose a chart neighbourhood $U_p \subset M$ of p such that

$$\varphi : U_p \to \varphi(U_p) \subset \mathbb{R}^n,$$

$\varphi(p) = 0$ and $\varphi = \exp_p^{-1}$. Here the tangent space T_pM is identified with $\mathbb{R}^n$ with the help of the derivative $\mathrm{d}\varphi(p)$. Thus the chart (U_p, φ) gives normal coordinates in a neighbourhood of the point p. Denote

$$f : U_p \times \mathbb{R}^n \to \mathbb{R},$$
$$f(x,y) = (g_{ij}(x)y^iy^j)^{1/2} = ||y||_g,$$

when $y \in T_xM$ and $x \in U_p$.

In normal coordinates $g_{ij}(p) = \delta_{ij}$ (Theorem 2.3). Therefore, for any $y \in T_pM$,

$$f(p,y) = (\sum_i (y^i)^2)^{1/2}.$$

Thus $||y||_g = |y|$, when $|\cdot|$ is the euclidean norm and $y \in T_pM$.

Choose $\epsilon > 0$ and a neighbourhood $U_p^\epsilon \subset U_p$ such that

$$\frac{1}{L}|y| \leq ||y||_g \leq L|y|,$$

for any $x \in U_p^\epsilon$ and $y \in T_xM$, when $L = 1 + \epsilon$. Denote $\mathbb{S}^{n-1} = \{y \in \mathbb{R}^n : |y| = 1\}$. Let us consider a restriction $f|_{U_p \times \mathbb{S}^{n-1}}$ of the function $f : U_p \times \mathbb{R}^n \to \mathbb{R}$. Choose $y \in \mathbb{S}^{n-1}$ and define a mapping

$$\tilde{f} : x \mapsto f^2(x,y) = (g_{ij}(x)y^iy^j).$$

This is a smooth mapping in the neighbourhood U_p, and, therefore, the mapping $h = \tilde{f} \circ \varphi^{-1} : \varphi(U_p) \to \mathbb{R}$ is smooth in the neighbourhood $\varphi(U_p)$. At the origin

$$h(0) = \tilde{f}(\varphi^{-1}(0)) = \tilde{f}(p) = |y|^2 = 1.$$

Denote $\epsilon' = \frac{\epsilon}{1+\epsilon}$ and let $u \in \varphi(U_p)$ be such a point that

$$|h(u) - h(0)| < \epsilon'.$$

For any such u

$$\frac{1}{L} = \frac{1}{1+\epsilon} = 1 - \epsilon' \leq h(u) \leq 1 + \epsilon' \leq 1 + \epsilon = L.$$

Denote the corresponding set of points $x = \varphi^{-1}(u)$ on the manifold as U_p^ϵ. Thus, for any $x \in U_p^\epsilon$ we have

$$\frac{1}{L} \leq \tilde{f}(x) \leq L.$$

We now have a neighbourhood U_p^ϵ such that for any $x \in U_p^\epsilon$

$$\frac{1}{L}|y| = \frac{1}{L} \leq f(x,y) = ||y||_g \leq L = L|y|, \tag{5}$$

where $y \in T_xM$, $|y| = 1$. Since the functions $y \mapsto |y|$ and $y \mapsto f(x,y)$ are 1-homogeneous, the equation (5) is true for any $y \in T_xM$ regardless of the norm of the vector y.

Let $\tilde{U}_p$ be such a neighbourhood of the point $p \in M$ that $\tilde{U}_p \subset U_p^\epsilon$ and $\varphi(\tilde{U}_p)$ is convex. Let x and y be points in $\tilde{U}_p$ and let $t \mapsto \gamma(t)$, $t \in [a,b]$, be such a path between x and y that its image under φ is a line segment between $\varphi(x)$ and $\varphi(y)$. We parametrize this path by curve length. Such a path can be chosen since $\varphi(\tilde{U}_p)$ is convex. Using the equation (5) and the definition of curve length, we get

$$\begin{aligned} d(x,y) &\leq l(\gamma) \\ &= \int_a^b \Big(g_{ij}(\varphi(\gamma(t))) \frac{\mathrm{d}\varphi(\gamma(t))^i}{\mathrm{d}t} \frac{\mathrm{d}\varphi(\gamma(t))^j}{\mathrm{d}t} \Big)^{1/2} \mathrm{d}t \\ &\leq L \int_a^b \Big| \frac{\mathrm{d}\varphi(\gamma(t))}{\mathrm{d}t} \Big| \,\mathrm{d}t \\ &= L|b-a| \frac{|\varphi(\gamma(b)) - \varphi(\gamma(a))|}{|b-a|} \\ &= L|\varphi(y) - \varphi(x)|, \end{aligned}$$

where

$$\Big| \frac{\mathrm{d}\varphi(\gamma(t))}{\mathrm{d}t} \Big| = \frac{|\varphi(\gamma(b)) - \varphi(\gamma(a))|}{|b-a|},$$

since $\varphi(\gamma(t))$ is a line.

According to Theorem 2.6, the point p has a geodetically convex neighbourhood U'_p. Let us now choose U'_p such that $U'_p \subset \tilde{U}_p$. Choose x, $y \in U'_p$. Let γ be a geodesic between x and y. Since U'_p is geodetically convex, the geodesic γ stays inside U'_p, and we have

$$\begin{aligned} d(x,y) &= l(\gamma) \\ &= \int_a^b \Big(g_{ij}(\varphi(\gamma(t))) \frac{\mathrm{d}\varphi(\gamma(t))^i}{\mathrm{d}t} \frac{\mathrm{d}\varphi(\gamma(t))^j}{\mathrm{d}t} \Big)^{1/2} \mathrm{d}t \\ &\geq \frac{1}{L} \int_a^b \Big| \frac{\mathrm{d}\varphi(\gamma(t))}{\mathrm{d}t} \Big| \,\mathrm{d}t \\ &\geq \frac{1}{L} |b-a| \frac{|\varphi(\gamma(b)) - \varphi(\gamma(a))|}{|b-a|} \\ &= \frac{1}{L} |\varphi(y) - \varphi(x)|. \end{aligned}$$

Combining these results, we get

$$\frac{1}{L} |\varphi(y) - \varphi(x)| \leq d(x,y) \leq L|\varphi(y) - \varphi(x)| \tag{6}$$

for any x, $y \in U'_p$, when $L = 1 + \epsilon$, $\epsilon > 0$.

Corollary 2.11. *For any riemannian manifold, such an atlas can be chosen that the coordinate charts are locally bilipschitz-continuous.*

If we consider a compact riemannian manifold, a global bilipschitz constant can be chosen: for a compact manifold we get an infinite cover from the chart neighbourhoods. Thus there is only a finite number of coordinate charts, and the greatest bilipschitz constant serves for all of them.

Corollary 2.12. *For a compact riemannian manifold, such an atlas can be chosen that all the coordinate charts are bilipschitz-continuous with a constant* $L > 1$.

3. DEFINITIONS AND PROPERTIES OF QUASIREGULAR MAPPINGS

In real spaces, when dimension $n \geq 3$, all conformal mappings are Möbius mappings [Ge], [Re]. Nevertheless, there exist mappings of the space $\mathbb{S}^n$ ($n \geq 3$) that modify the euclidean structure in a controlled manner and preserve certain conformic characteristics. They have many features in common with holomorphic mappings, for example, the famous theorems by Picard and Montel. These mappings are quasiregular mappings. Their dynamics can be studied with the help of a certain subgroup called uniformly quasiregular mappings.

At the beginning of this chapter, we define quasiregular and uniformly quasiregular mappings, first in euclidean spaces and then, more generally, on riemannian manifolds. We also define branch sets and normal families and consider some of their properties.

3.1. **Quasiregular mappings.**

There are various equivalent ways of defining quasiregular mappings, depending on whether the desired point of view is more geometric, topological or analytic. We choose the following definition.

Definition 3.1. Let $D \subset \overline{\mathbb{R}}^n$ be a domain and $f : D \to \overline{\mathbb{R}}^n$ a non-constant mapping of the Sobolev class $W^{1,n}_{loc}(D)$. We consider only orientation-preserving mappings, which means that the Jacobian determinant $J_f(x) \geq 0$ for a.e. $x \in D$. Such a mapping is said to be *K-quasiregular*, where $1 \leq K < \infty$, if

$$\max_{|h|=1} |f'(x)h| \leq K \min_{|h|=1} |f'(x)h|$$

for a.e. $x \in D$, when f' is the formal matrix of weak derivatives. The smallest number K for which the above inequality holds is called the *linear dilatation*.

We can generalize this definition to riemannian manifolds with the help of bilipschitz-continuous coordinate charts. From now on, we expect all manifolds to be smooth, unless otherwise stated.

Definition 3.2. Let M and N be n-dimensional riemannian manifolds. A non-constant continuous mapping $f : M \to N$ is K-quasiregular if for every $\varepsilon > 0$ and every $m \in M$ there exists bilipschitz-continuous charts (U, φ), $m \in U$, and (V, ψ), $f(m) \in V$, so that the mapping $\psi \circ f \circ \varphi^{-1}$ is $(K + \varepsilon)$-quasiregular (see Figure 1).

Later we will consider only compact manifolds. Then we can choose coordinate atlas where the coordinate mappings all have the same bilipschitz constant $L = 1+\varepsilon$, $\varepsilon > 0$ (see Corollary 2.12). Thus we see that on compact manifolds Definition 3.2 is a global one.

A non-constant quasiregular mapping can be redefined in a set of measure zero such that the mapping is made continuous, open and discrete [R2]. We will henceforth assume that quasiregular mappings always have these properties.

Lemma 3.3. *Let M and N be compact n-dimensional riemannian manifolds and $f : M \to N$ a quasiregular mapping. Then the set $\{f^{-1}(p)\}$ of preimage points is finite for every $p \in N$.*

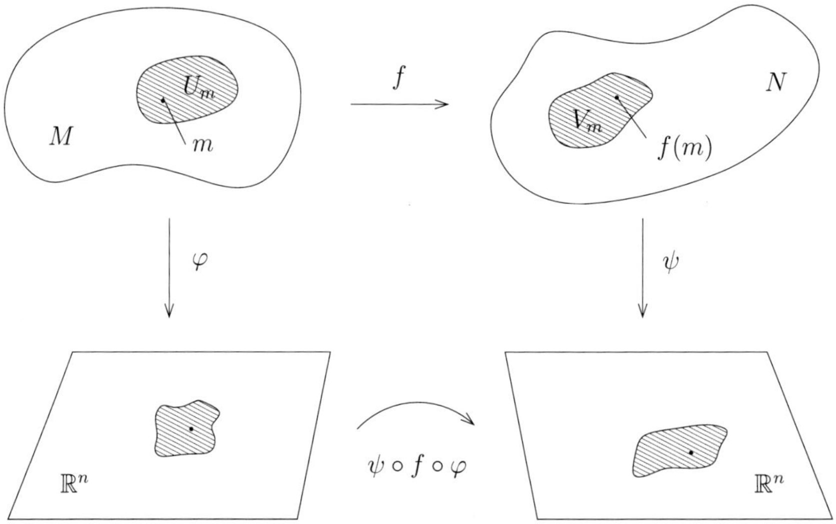

FIGURE 1. Quasiregular mapping between riemannian manifolds.

Proof. Since f is quasiregular, it is discrete, which means that the connected components of $\{f^{-1}(p)\}$ are singletons for every $p \in N$. Since M is compact, the set $\{f^{-1}(p)\}$ must then be finite: if it was infinite, it would contain an accumulation point by Bolzano–Weierstrass theorem, which would be in contradiction with the discreteness.

□

3.2. **Branch set.** The *branch set* B_f of the mapping $f : M \to N$ is the set of those points $x \in M$ where f is not locally homeomorphic. In other words, $x \in B_f$ if and only if for all open neighbourhoods U of x the restricted mapping $f|U$ is not injective. Notice that

$$B_{f^2} = B_f \cup f^{-1}(B_f)$$

and, more generally,

$$B_{f^n} = \cup_{i=0}^{n-1}(f^i)^{-1}(B_f),\ n \in \mathbb{Z}_+.$$

Quasiregular homeomorphisms are called *quasiconformal* (abbreviated qc) mappings. For qc-mappings $B_f = \emptyset$, since they are homeomorphisms.

Theorem 3.4. *For the branch set B_f of the quasiregular mapping $f : M \to N$ between riemannian n-manifolds M and N, we have*

$$\dim(B_f) = \dim f(B_f) \leq n - 2.$$

This (as many other results considering the branch set) holds more generally for any open and discrete mapping between n-manifolds. A.V. Chernavskiĭ was the first to prove this theorem in [Ch1], [Ch2], but a more illustrative proof can be found in the article [V1].

At the moment, it is not known whether there exist quasiregular mappings with a non-empty branch set of dimension less than $n-2$. Such open and discrete mappings are, however, known, see, for example, [CT], where P. T. Church and J. G. Timourian present a mapping f in dimension $n = 5$ with $B_f \neq \emptyset$ and $\dim(B_f) < 3$.

Let us now consider the branch set of a quasiregular mapping more closely. Let $f : M \to M$ be a quasiregular mapping on a riemannian manifold M. Outside the branch set f is a local homeomorphism. If we assume that $B_f = \emptyset$, the mapping is a local homeomorphism everywhere. On the other hand, we know that since a quasiregular local homeomorphism is a covering map, it is a homeomorphism on a simply connected manifold [V2, Theorem 24.10]. If the manifold is not simply connected, the emptyness of the branch set does not guarantee that the mapping is homeomorphic (for such an example, see [Ka, p. 21]).

G. Martin, V. Mayer and K. Peltonen have proved that closed manifolds which have a non-injective uniformly quasiregular mapping with an empty branch set, are so-called flat manifolds [MMP]. This means that they can be obtained as quasiregular images of spaces of the form $\mathbb{R}^n/\Gamma$, where Γ is a Bieberbach group. For that reason, in this thesis we will consider only mappings with non-empty branch sets.

3.3. **Uniformly quasiregular mappings.** Let us now define uniformly quasiregular mappings on compact riemannian manifolds.

Definition 3.5. Let M be a compact riemannian manifold. A non-injective mapping f from a domain $D \subset M$ onto itself is called *uniformly quasiregular* (uqr) if there exists a constant $1 \leq K \leq \infty$ such that all the iterates f^k are K-quasiregular.

The set of uniformly quasiregular mappings on the domain D is denoted by $UQR(D)$. Let us point out that a uniformly quasiregular mapping $f : M \to M$ on a compact riemannian manifold M is necessarily a surjection, since quasiregular mappings are continuous and open. Thus the image fM is both compact and open at the same time, which means that $fM = M$. In addition, we will assume our uniformly quasiregular mappings to be non-injective.

In dimension $n = 2$, uniformly quasiregular mappings exist on the sphere $\mathbb{S}^2$ and on the torus T^2. On the sphere, all rational functions are uqr-mappings. On the other hand, all uqr-mappings of $\mathbb{S}^2$ are, up to quasiconformal change of coordinates, rational functions [Ma2, p. 21]. Thus the theory of uniformly quasiregular mappings on $\mathbb{S}^2$ is the theory of rational functions.

On 2-dimensional manifolds, which have genus 2 or greater, uniformly quasiregular mappings do not exist. This follows from A. W. Tucker's results in [Tu]. Tucker proves that for the so-called simplicial mappings $s : M \to N$ between manifolds M and N, we have

$$\chi(M) + \sum_{x \in B_s} (i(x,s) - 1) = \chi(N)d, \tag{7}$$

where $\chi(M)$ is the Euler characteristic of the manifold M, $i(x,s)$ is the local degree of the mapping s at the point x, and d is the degree of the mapping s. In general, quasiregular mappings are not simplicial, but in dimension 2, uniformly quasiregular mappings are simplicial up to quasiconformal conjugation. The reason for this is that any 2-manifold of genus at least 2 can be conformally covered by the disc D^2, and so a uqr-mapping $f : M^2 \to M^2$ can be lifted conformally to a uqr-mapping $\tilde{f} : D^2 \to D^2$ on the disc. Any uqr-mapping on D^2 is a rational mapping up to a quasiconformal change of variables [Ma2, p. 21].

Let M_g^2 be 2-manifold with genus g. For its Euler characteristic we have $\chi(M_g^2) = 2 - 2g$ [Lee, p. 169]. Thus for any simplicial mapping $s : M_g^2 \to M_g^2$, it follows from the equation (7) that

$$\sum_{x \in B_s} (i(x,s) - 1) = (d-1)(2-2g). \tag{8}$$

The left side of this equation is never negative, and so we must have $g \leq 1$. Thus, on 2-dimensional manifolds with genus greater than 1, any simplicial mapping is a homeomorphism. If $g = 1$, we see from the equation (8) that $i(x,s) = 1$ for any x. Thus, on the torus, all uniformly quasiregular mappings are locally homeomorphic. The genus of the 2-sphere is zero. So in the case of the sphere, the equation (8) tells us only that the degree of the mapping is greater when there is more branching.

When the dimension is $n \geq 3$, the theory of quasiregular mappings differs a lot from the 2-dimensional case. After Liouville's theorem in $\mathbb{R}^n$, any 1-quasiregular mapping is either constant or a restriction of a Möbius mapping [R2, p. 11]. It is also known that for every $n \geq 3$ there exists a constant $K_0 > 1$ such that any K_0-quasiregular mapping is a local homeomorphism, but the constant K_0 is not yet known [R2, p. 161]. With the following two theorems, T. Iwaniec and G. J. Martin have proved that for $K > 2$ there do exist non-injective uniformly K-quasiregular mappings also in dimensions $n \geq 3$ [IM1, Theorems 21.2.1 and 21.2.2]. (Denote that the values of the constant K may differ in the literature depending on which of the equivalent definitions is chosen.)

Theorem 3.6. *For every $K > 2$ there is an infinite K-quasiregular semigroup Γ acting on $\overline{\mathbb{R}}^n$ with the property that every element of Γ has a non-empty branch set.*

Theorem 3.7. *Let $f : \overline{\mathbb{R}}^n \to \overline{\mathbb{R}}^n$ be a quasiregular mapping with branch set B_f. Then there is a uniformly quasiregular mapping $g : \overline{\mathbb{R}}^n \to \overline{\mathbb{R}}^n$ with $B_g = B_f$.*

A *K-quasiregular semigroup* is a family of K-quasiregular mappings, which is closed under the composition of the mappings. Thus, at least in $\overline{\mathbb{R}}^n$, there exists a uniformly K-quasiregular mapping with non-empty branch set for every $K > 2$. It has been suggested that the distortion K is greater than 2 for any branching quasiregular mapping in $\overline{\mathbb{R}}^n$, $n \geq 3$. If this holds, Iwaniec and Martin have given an optimal answer for the question of existence of uqr-mappings in $\overline{\mathbb{R}}^n$. Proofs of these theorems are based on the so-called conformal trap method, which they have developed.

With a generalization of the same trap method, uniformly quasiregular mappings can in fact be constructed on any smooth riemannian manifold M^n that has $\mathbb{S}^n$ as a universal covering space [Pe, Theorem 4]. Thus uqr-mappings exist, for example, on three-dimensional lens spaces and Poincaré homology spheres.

3.4. **Lattès-type mappings.** In 1997, V. Mayer discovered an important family of examples of uniformly quasiregular mappings (see [Ma1]). He generalized Lattès' construction of so-called chaotic rational maps. The resulting chaotic uqr-mappings of $\overline{\mathbb{R}}^n$ are constructed as follows:

Definition 3.8. Let Υ be a discrete group of isometries of $\mathbb{R}^n$. A mapping $h : \mathbb{R}^n \to M$ is *automorphic* with respect to Υ in the *strong sense* if

(1) $h \circ \gamma = h$ for any $\gamma \in \Upsilon$,
(2) Υ acts transitively on the fibres $\mathcal{O}_y = h^{-1}(y)$.

By the latter condition we mean that for any two points x_1, x_2 with $h(x_1) = h(x_2)$ there is an isometry $\gamma \in \Upsilon$ such that $x_2 = \gamma(x_1)$. We have the following theorem by T. Iwaniec and G. Martin [IM1, pp. 501-502]. The proof in [IM1] is written for the case $M = \mathbb{R}^n$, but it holds more generally to any riemannian manifold M without changes.

Theorem 3.9. *Let Υ be a discrete group such that $h : \mathbb{R}^n \to M$ is automorphic with respect to Υ in the strong sense. If there is a similarity $A = \lambda\mathcal{O}$, $\lambda \in \mathbb{R}$, $\lambda \neq 0$, and $\mathcal{O}$ an orthogonal transformation, such that*

$$A\Upsilon A^{-1} \subset \Upsilon,$$

then there is a unique solution $f : h(\mathbb{R}^n) \to h(\mathbb{R}^n)$ to the Schröder functional equation

$$f \circ h = h \circ A$$

and f is a uniformly quasiregular mapping.

By the condition $A\Upsilon A^{-1} \subset \Upsilon$ it is simply meant that for any $\gamma \in \Upsilon$ there exists a $\gamma' \in \Upsilon$ such that $A\gamma(x) = \gamma' A(x)$ for any x. The idea of the proof presented in [IM1] is that if h is automorphic with respect to a discrete group, it does not recognize whether the space has been "moved" or not. If, in addition, A does not disturb the action of the group, there is a solution to the Schröder equation. Note that we now have also the equation $f^k \circ h = h \circ A^k$ for all k.

3.5. **Normal family.** When iterating quasiregular mappings and considering their Julia sets, we will need the concept of normal families. These families and their properties are discussed in detail in [Sch]. To get started with the definitions, let $\mathcal{F}$ be a family of mappings f from a topological space X into a metric space (Y, μ).

Definition 3.10. The family $\mathcal{F}$ is *equicontinuous at the point* $a \in X$ if for any $\varepsilon > 0$ there exists a neighbourhood U of the point a such that

$$\mu(f(x), f(a)) < \varepsilon$$

for any $x \in U$, $f \in \mathcal{F}$. The family $\mathcal{F}$ is *equicontinuous* if it is equicontinuous at every point of the space X.

Definition 3.11. The family $\mathcal{F}$ is *normal* if every sequence $(f_j)_{j=1}^{\infty}$ of its mappings has a subsequence (f_{j_k}), which is locally uniformly converging in X, that is, uniformly converging on every compact subset of X. A normal family which contains its limit mappings is *closed.*

The following theorem combines these properties. It is proved, for example, in [IM1, p. 480].

Theorem 3.12 (Arzela—Ascoli). *An equicontinuous family $\mathcal{F}$ of mappings from a separable topological space X to a metric space Y is normal provided that the closure of the set $\{f(x) : f \in \mathcal{F}\}$ is compact in Y for each $x \in X$.*

Lemma 3.13. *Let $\mathcal{F}$ be a family of quasiregular mappings $f : M \to N$ between two riemannian manifolds M and N. If for every $m \in M$ there exists a neighbourhood $U \subset M$ such that the family*

$$\mathcal{F}|_U := \{f|_U : U \to N \mid f \in \mathcal{F}\}$$

is normal, then also the family $\mathcal{F}$ is normal.

Proof. Let (f_j) be an arbitrary sequence of mappings of the family $\mathcal{F}$. We will prove with the so-called diagonal argument that the sequence (f_j) has a subsequence, which converges locally uniformly on the manifold M.

Since the second countability axiom holds for manifolds by definition, all manifolds are so-called Lindelöf spaces [V2, p. 91]. It means that every open cover has a countable subcover. Hence we can choose for M a countable cover $\{U_j\}$, which consists of some neighbourhoods U_j, such that the family $\mathcal{F}$ is normal in each of them.

There exists a subsequence (f_{j1}) of our arbitrarily chosen sequence (f_j), which converges locally uniformly in the neighbourhood U_1. The sequence (f_{j1}) belongs to the family $\mathcal{F}$, and therefore it has a locally uniformly converging subsequence in the neighbourhood U_2. Let us denote this subsequence (f_{j2}). Still $(f_{j2}) \subset \mathcal{F}$, and it has a subsequence (f_{j3}) which converges locally in U_3. If we continue in the same manner, we get for every U_k a locally uniformly converging sequence $(f_{jk}) \in \mathcal{F}$, which is a subsequence of all the previous sequences (f_{jl}), $l < k$.

Let us now define a new sequence (f_l) as follows: we take the first mapping of the sequence (f_{j1}) as the first element of our new sequence. For the second element, we take the second mapping from (f_{j2}), for the third element the third mapping from (f_{j3}), and so on. Thus the lth element of the sequence (f_l) is the lth mapping of the subsequence (f_{jl}). This new sequence (f_l) now converges locally uniformly in every neighbourhood U_j and so also in their union, which is the whole manifold M. Therefore, the family $\mathcal{F}$ is normal on the manifold M. □

4. DYNAMICS OF UNIFORMLY QUASIREGULAR MAPPINGS

In this chapter we consider the dynamics of uniformly quasiregular mappings. We define their Fatou and Julia sets and prove that the Julia set is always non-empty. To be able to consider the Julia and Fatou sets and their properties, we have to define what we mean by normality of sequences of uniformly quasiregular mappings on manifolds in terms of local representations. This allows us to apply known results for qr-mappings in euclidean spaces. As we pointed out earlier in chapter 3, all uniformly quasiregular mappings $f : M \to M$ are surjections if M is compact. In addition, we assume them to be non-injective. Consequently, constant mappings and identity mappings are ruled out from our considerations.

4.1. **Local representations.** Let M be again a compact riemannian manifold and $f : M \to M$ a uniformly quasiregular mapping. The mapping f can be iterated without problems on M, but we do not know anything about the convergence of the sequence $(f^k)_{k=1}^{\infty}$ of the iterates. So first we have to consider the local behaviour of the mappings to find out what happens globally.

Let us consider an arbitrary point $m \in M$. We wish to define a local representation for any iterate f^k near this point m. Especially, we want this local representation to be a uqr-mapping.

In chapter 2 we proved that for a compact riemannian manifold such an atlas can be chosen that all the coordinate charts are bilipschitz-continuous with the same constant L. Let now $(\varphi_m, U_{\varphi_m})$ be such an L-bilipschitz chart near the point m that $\varphi_m(m) = 0$. Let $\{(\psi'_\alpha, U_\alpha) : \alpha \in I\}$ be such a collection of finitely many L-bilipschitz charts that $\cup_\alpha U_\alpha = M$. A finite collection is enough, since M is compact. With this collection and translations we construct an atlas $\mathcal{A}$ for the manifold M as follows. Let U_k be that coordinate neighbourhood for which $f^k(m) \in U_k$. Let ψ_k be a mapping defined in U_k such that

$$\psi_k := t_k \circ \psi'_{\alpha_k},$$

where t_k is a translation, $t_k(x) = x - \psi'_{\alpha_k}(f^k(m))$. Then,

$$\psi_k(f^k(m)) = 0,$$

and we get an atlas $\mathcal{A} = \{(\psi_k, U_k) : k = 1 \ldots \infty\}$ of L-bilipschitz mappings with only finitely many different coordinate neighbourhoods. Also the set of coordinate charts is finite up to translations. The mapping φ_m is used as a coordinate chart near m and ψ_k near $f^k(m)$, and they map exactly these points to origin.

Let us now define a local representation for the iterate f^k near the point m. Let $U \subset U_{\varphi_m}$ be an open neighbourhood near the point m, such that $f(U) \subset U_1$ (see Figure 2).

The mapping

$$g_1 := \psi_1 \circ f \circ \varphi_m^{-1}$$

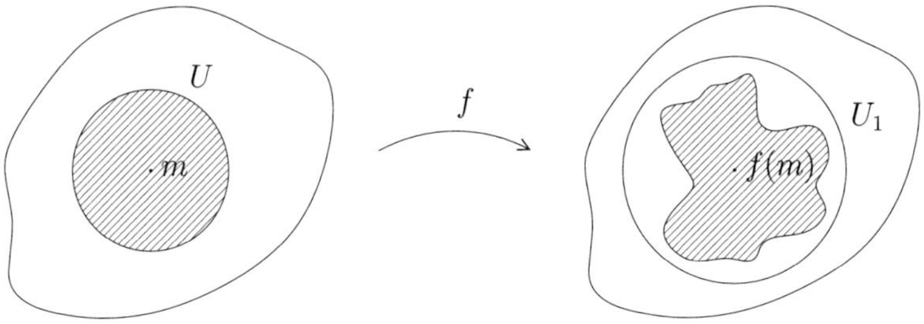

FIGURE 2. The neighbourhood U is mapped into U_1.

is defined in the domain $\varphi_m(U) \subset \mathbb{R}^n$, and it is K'-quasiregular. The constant K' depends on the dilatation K of the mapping f and the bilipschitz constant L of the atlas $\mathcal{A}$.

For f^2 we wish to define a mapping g_2 in the same domain $\varphi_m(U) \subset \mathbb{R}^n$. For this purpose we need a certain mapping to scale the domain: For any $k \geq 2$ choose a neighbourhood $W_k \subset U_{\varphi_m}$ of m such that $V_k := f^k(W_k) \subset U_k$. Though W_k belongs to the set U_{φ_m}, the set $\{f^{-k}(V_k)\}$ may well have other components outside the set W_k. Lemma 4.1 shows that in a neighbourhood $\tilde{U}_m \subset U$ of m we can define a conformal scaling map $s_k : \mathbb{R}^n \to \mathbb{R}^n$ such that

$$s_k(\varphi_m(\tilde{U}_m)) \subset \varphi_m(W_k)$$

as $s_k : x \mapsto a_k x$, where a_k is small enough. If we then define

$$g_k := s_k^{-1} \circ \psi_k \circ f^k \circ \varphi_m^{-1} \circ s_k,$$

the mapping g_k is defined in the same domain $\varphi_m(\tilde{U}_m)$ for any k and it is K'-quasiregular, since the coordinate charts are L-bilipschitz-continuous and the scaling map is conformal.

Lemma 4.1. *There exist constants $R_k > 0$ and $r_k > 0$ such that we can choose the scaling map s_k to be*

$$s_k : x \mapsto \frac{r_k}{R_k} x$$

for any $k \geq 2$.

Proof. Let m be an arbitrary point on M. Let $\rho = \rho(m) > 0$ be such a constant that for any point $x \in M$ the ball $\mathcal{B}(x, \rho)$ belongs to some coordinate neighbourhood of the atlas $\{(\psi_k, U_k)\}$, which we defined earlier. The constant ρ can be found since there are only a finite number of coordinate neighbourhoods in this atlas.

Choose a neighbourhood $V_{k,m} = \mathcal{B}(f^k(m), \rho) \subset U_k$ of the point $f^k(m)$. Let $W_{k,m} \subset U_{\varphi_m}$ be that component of the set $f^{-k}(V_{k,m})$ which contains $m \in U_{\varphi_m}$ (see Figure 3). If the set $W_{k,m}$ is not inside U_{φ_m}, we take $W_{k,m} \cap U_{\varphi_m}$ instead of $W_{k,m}$.

Since the set $\varphi_m(W_{k,m})$ is open, there exists a constant $r_k > 0$ such that

$$\mathbb{B}(\varphi_m(m), r_k) \subset \varphi_m(W_{k,m}).$$

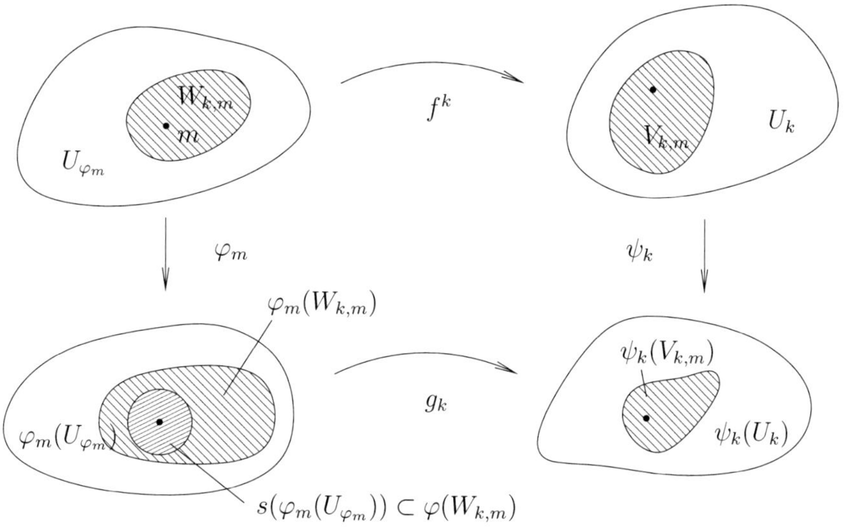

FIGURE 3. Some neighbourhoods and mappings needed in the proof of Lemma 4.1.

Define $\tilde{U}_m = U_{\varphi_m}$. The set $\varphi_m(\tilde{U}_m)$ is bounded, because the mapping φ_m is bilipschitz-continuous and the set $\tilde{U}_m$ is bounded. Therefore we can define another constant $R_k > 0$ as $R_k = 2 \operatorname{diam} (\varphi_m(\tilde{U}_m))$.

Now we can choose for any k the constant of the scaling map to be $a_k = \frac{r_k}{R_k}$. Thus, $s_k : x \mapsto \frac{r_k}{R_k} x$ and

$$s_k : \varphi_m(\tilde{U}_m) \subset \mathbb{B}(\varphi_m(m), R_k) \to \mathbb{B}(\varphi_m(m), r_k) \subset \varphi_m(W_{k,m}).$$

□

From now on we choose our scalings s_k according to Lemma 4.1 if we need scaling and the previous scaling s_{k-1} is not enough. Thus we always define $s_k = \text{id}$ if possible, and if scaling is necessary, $s_k = s_{k-1}$ if the previous scaling is strong enough also for this iterate. If it is not, we use $s_k : x \mapsto \frac{r_k}{R_k} x$ with R_k and r_k as in Lemma 4.1. Especially this means that if the sequence (f^k) converges, after some index k_0 all the scalings are the same: $s_k = s_{k_0}$ for all $k \geq k_0$.

Note that in addition to k, the scaling map s_k depends also on the base point m. This does not matter, since we were anyway defining local representations, that is something that strongly depends on the point which one is considering. We will from now on notify the point as a subscript. Thus the mappings of the local representation at the point m are denoted by

$$g_{m,k} := s_{m,k}^{-1} \circ \psi_{m,k} \circ f^k \circ \varphi_m^{-1} \circ s_{m,k} : \varphi_m(\tilde{U}_m) \to s_{m,k}^{-1} \circ \psi_{m,k}(U_{m,k}).$$

The sequence $(g_{m,k})_{k=1}^{\infty}$ is what we call *a local representation of the iterates* $(f^k)_{k=1}^{\infty}$ *at the point* m, and the K'-quasiregular mappings $g_{m,k}$ are defined in the neighbourhood $\varphi_m(\tilde{U}_m)$ of the point $\varphi_m(m)$.

Remark 4.2. *Local representations are not unique. The coordinate charts and scalings which are needed in defining the representation can be chosen freely as long as they fulfil the demands illustrated earlier.*

Definition 4.3. The local representation $(g_{m,k})_{k=1}^{\infty}$ is *normal* at $\varphi_m(m)$ if for any sequence (g_{m,l_i}) of the family $\{(g_{m,l_i}) \mid l_i < l_{i+1} \forall i \in \mathbb{N}\}$ there exists a subsequence $(g_{m,l_{i_j}})$ which converges locally uniformly in a neighbourhood of $\varphi_m(m)$ towards a limit mapping G_m.

This limit mapping G_m is a mapping from a subset of the set $\varphi_m(\tilde{U}_m)$, where all the mappings $g_{m,k}$ are defined and the sequence $(g_{m,l_{i_j}})$ converges locally uniformly, into the set

$$U_{G_m} := \cap_k s_{m,k}^{-1} \circ \psi_{m,k}(U_{m,k}).$$

This set cannot be just one point, since the mappings $\psi_{m,k}$ are bilipschitz-continuous and the sequence (a_k) is bounded from above. Therefore it contains some ball $\mathbb{B}(0, \rho)$, where $\rho > 0$.

4.2. **Julia sets.** In chapter 3 we defined a K-quasiregular semigroup to be a family of K-quasiregular mappings, which is closed under composition of the mappings. Now we can define Julia and Fatou sets of such semigroups. We present first this global definition, but in fact we are interested in considering the local behaviour with the help of the local representations defined in the previous chapter.

Definition 4.4. Let Γ be a quasiregular semigroup. Then the *Fatou set* of Γ is

$$\mathcal{F}(\Gamma) = \{x \in M : \quad \text{there exists an open set } U \subset M \text{ such that } x \in U \text{ and } \Gamma|U \text{ is normal}\}.$$

The *Julia set* of the family Γ is

$$\mathcal{J}(\Gamma) = M \setminus \mathcal{F}(\Gamma).$$

By the definition, Fatou sets are open, and therefore Julia sets are closed. If the family Γ consists of iterates of a mapping f, that is $\Gamma = \{f^k \mid k = 1, 2, \ldots\}$, we call these sets simply Fatou and Julia sets of the mapping f, denoted by $\mathcal{F}_f$ and $\mathcal{J}_f$.

Lemma 4.5. *Both Fatou and Julia set are completely invariant, in other words, for any* $f \in \Gamma$ *we have*

$$f(\mathcal{F}(\Gamma)) = f^{-1}(\mathcal{F}(\Gamma)) = \mathcal{F}(\Gamma),$$
$$f(\mathcal{J}(\Gamma)) = f^{-1}(\mathcal{J}(\Gamma)) = \mathcal{J}(\Gamma).$$

Proof. To simplify notations, we prove the lemma for Fatou and Julia sets of a quasiregular mapping f. Since $\mathcal{J}_f = M \setminus \mathcal{F}_f$ and f is surjective, it is sufficient to prove that $\mathcal{J}_f$ is completely invariant regarding to f.

Let $z \in \mathcal{J}_f$ and $w = f(z) \in f(\mathcal{J}_f)$. Assume that $w \notin \mathcal{J}_f$. Then $\{f^n\}$ is normal in a neighbourhood of w, and it follows that $\{f^{n+1}\}$ is normal in some neighbourhood of z because f is continuous. This is a contradiction, and therefore $w \in \mathcal{J}_f$. We have now proved the inclusion $f(\mathcal{J}_f) \subset \mathcal{J}_f$.

Let y be such a point that $f(y) = z \in \mathcal{J}_f$. Assume that $y \notin \mathcal{J}_f$, meaning that $\{f^n\}$ is normal in a neighbourhood of y. Since f is open, $\{f^{n-1}\}$ is normal in some neighbourhood of z, which is again a contradiction to $z \in \mathcal{J}_f$. Thus $y \in \mathcal{J}_f$, that is $f^{-1}(\mathcal{J}_f) \subset \mathcal{J}_f$.

We now know that $f(\mathcal{J}_f) \subset \mathcal{J}_f$ and $f^{-1}(\mathcal{J}_f) \subset \mathcal{J}_f$. Applying f^{-1} to the former inclusion and f to the latter and noting that $\mathcal{J}_f \subset f^{-1}(f(\mathcal{J}_f))$, we see that

$$f(\mathcal{J}_f) = f^{-1}(\mathcal{J}_f) = \mathcal{J}_f.$$

□

To be able to work with Julia sets on manifolds, we have to define what we mean by normality. Definition 4.4 alone is not useful. We need now the local representations defined in Chapter 4.1.

Definition 4.6. Let $f : M \to M$ be a uniformly quasiregular mapping on a compact riemannian manifold M. The sequence of iterates $(f^k)_{k=1}^{\infty}$ is representationally normal at the point $m \in M$ if the local representation $(g_{m,k})_{k=1}^{\infty}$ is normal at $\varphi_m(m)$.

To make sure that the concept of representational convergence is well defined, we need to show that the convergence does not depend on the choice of the local representation. Assume that $(g_{m,k})_{k=1}^{\infty}$ and $(\tilde{g}_{m,k})_{k=1}^{\infty}$ are two local representations for a uqr-mapping f, and assume that $(g_{m,k})_{k=1}^{\infty}$ is normal at $\varphi_m(m) = 0$. We restrict our considerations to a neighbourhood of the origin, say U_m, where both local representations are defined. In this neighbourhood we may take the same scaling maps s_k for both mappings.

Since $(g_{m,k})$ is normal, there is a converging subsequence (g_{m,k_j}). After some index j_0, a subset of the neighbourhood U_m is mapped into one chart neighbourhood for all j, namely the same neighbourhood which contains the point $g(0)$. Thus all the chart mappings ψ_{m,k_j} are of the form $t_{m,k_j} \circ \psi$, when $j \geq j_0$.

$$g_{m,k_j} = s_{m,k}^{-1} \circ t_{m,k_j} \circ \psi \circ f^{k_j} \circ \varphi_m^{-1} \circ s_{m,k}.$$

This same must happen for the mappings $\tilde{g}_{m,k_j}$, since the mappings f^{k_j} take a neighbourhood of m into one chart neighbourhood. Thus we have also

$$\tilde{g}_{m,k_j} = s_{m,k}^{-1} \circ \tilde{t}_{m,k_j} \circ \tilde{\psi} \circ f^{k_j} \circ \tilde{\varphi}_m^{-1} \circ s_{m,k}.$$

We consider the smaller of the two neighbourhoods where both the above equations hold. Now

$$\tilde{g}_{m,k_j} = s_{m,k}^{-1} \circ \tilde{t}_{m,k_j} \circ (\tilde{\psi} \circ \psi^{-1}) \circ \psi \circ f^{k_j} \circ \varphi_m^{-1} \circ (\varphi_m \circ \tilde{\varphi}_m^{-1}) \circ s_{m,k}.$$

The mappings $\tilde{\psi} \circ \psi^{-1}$ and $\varphi_m \circ \tilde{\varphi}_m^{-1}$ are L^2-bilipschitz continuous homeomorphisms and the difference between the translations $\tilde{t}_{m,k_j}$ and t_{m,k_j} converges, since by Lemma 4.8 $f^{k_j} \to F$ for some limit mapping F, and

$$\begin{aligned} |t_{m,k_j}(x) - \tilde{t}_{m,k_j}(x)| &= |\psi(f^{k_j}(x)) - \tilde{\psi}(f^{k_j}(x))| \\ &\to |\psi(F(x)) - \tilde{\psi}(F(x))|. \end{aligned}$$

Therefore also the sequence $(\tilde{g}_{m,k_j})$ converges locally uniformly at $\varphi_m(m)$. Thus the convergence does not depend on the choice of the local representation.

Now we can formulate a local definition for Julia sets of uniformly quasiregular mappings on riemannian manifolds:

Definition 4.7. A point $m \in M$ belongs to the Julia set $\mathcal{J}_f$ of a uniformly quasiregular mapping $f : M \to M$ if the local representation $(g_{m,k})_{k=1}^{\infty}$ of the sequence $(f^k)_{k=1}^{\infty}$ of the iterates is not normal at the point $\varphi_m(m)$.

Our goal in this chapter is to prove that the Julia sets of uqr-mappings f with $\deg(f) \geq 2$ are non-empty. To obtain this result, we first prove two lemmas. The first of these lemmas also shows that the definitions 4.4 and 4.7 are in fact equivalent.

Lemma 4.8. *The sequence $(g_{m,k})$ has a subsequence (g_{m,k_j}) converging locally uniformly in $\varphi_m(\tilde{U}_m)$ to a mapping G_m if and only if there exists a subsequence (f^{k_u}) of (f^k) converging locally uniformly to a mapping F in a neighbourhood of the point m.*

Proof. Assume that $g_{m,k_j} \to G_m$ uniformly in $\varphi_m(\tilde{U}_m)$ when $j \to \infty$. (Otherwise take a smaller set where the convergence is uniform and denote its preimage by $\tilde{U}_m$.)

Recall the reason why we needed the scalings in the local representations was to make sure that the image after mapping with f^k ends up into one coordinate neighbourhood. We defined the coefficients a_k in $s_{m,k} : x \mapsto a_k x$ to be small enough to make sure that this happens, but not unnecessarily small. Especially, if f is contracting, the scalings after some index are equal. Thus the sequence (a_k) can converge to zero only when the mappings f^k blow up a neighbourhood of m. But also in this case the convergence of $(g_{m,k})$ tells us that after some index the image set $g_{m,k_j}(\varphi_m(\tilde{U}_m))$ is always in some bounded domain (near $G_m(\varphi_m(\tilde{U}_m))$), and we know that the coefficients a_k have not been smaller than some constant: For any $\delta > 0$ there is an index j_0 such that when $j \geq j_0$,

$$g_{m,k_j}(\varphi_m(\tilde{U}_m)) \subset \mathbb{B}(0, \rho + \delta),$$

where $\rho = \max\{|x| : x \in G_m(\varphi_m(\tilde{U}_m))\}$. This means that the scaling mappings s_{m,k_j} for $j \geq j_0$ can be chosen as $s_{m,k_j} = s_m$, where $s_m : x \mapsto ax$ and a is such a constant that $\mathbb{B}(0, a(\rho + \delta))$ fits inside the chart neighbourhood $\psi_{k_{j0}}(U_{k_{j0}})$.

Thus, when $j \geq j_0$, we have

$$g_{m,k_j} = s_m^{-1} \circ \psi_{m,k_j} \circ f^{k_j} \circ \varphi_m^{-1} \circ s_m.$$

Now we can take a subsequence (g_{m,k_l}) such that all the chart mappings are of the form $\psi_{m,k_l} = t_{m,k_l} \circ \psi$, since there was only a finite number of mappings ψ_k in the atlas.

We recall that the translation t_{m,k_l} takes the point $\psi \circ f^{k_j}(m)$ to the origin. If the set $\{f^{k_j}(m)\}$ has only finite number of points, some of the translations must occur infinetely many times in the sequence (g_{m,k_l}). Thus we can take such a subsequence (g_{m,k_u}) of the sequence (g_{m,k_l}) that all the translations are the same. Then

$$g_{m,k_u} = s_m^{-1} \circ t \circ \psi \circ f^{k_u} \circ \varphi_m^{-1} \circ s_m$$

and since (g_{m,k_u}) converges, also (f^{k_u}) must converge.

If the set $\{f^{k_j}(m)\}$, however, has infinitely many different points, then by Bolzano–Weierstrass theorem, it must also have an accumulation point p. Thus we can pick a subsequence $(f^{k_u}(m))$ of these points, which converges to p. Now when $u \to \infty$, the translations t_{m,k_u} become more and more alike: for any $\epsilon > 0$ there is an index n_ϵ such that when $i,\, j > n_\epsilon$,

$$|t_{m,k_i}(x) - t_{m,k_j}(x)| < \epsilon$$

for any x. Thus, since the sequence (g_{m,k_u}) of the mappings

$$g_{m,k_u} = s_m^{-1} \circ t_{m,k_u} \circ \psi \circ f^{k_u} \circ \varphi_m^{-1} \circ s_m$$

converges and the translations converge to $x \mapsto x - \psi(p)$, also the sequence (f^{k_u}) must converge uniformly in $\varphi_m^{-1} \circ s_m \circ \varphi_m(\tilde{U}_m)$.

Assume then that (f^{k_u}) converges locally uniformly to a mapping F in a neighbourhood of the point m. We construct a local representation that converges locally uniformly in some neighbourhood of $\varphi(m)$. Take some radius r such that $\mathbb{B}(F(m), r)$ belongs to one chart neighbourhood (U_ψ, ψ). Then we can find a neighbourhood W of m such that $f^{k_u}(W) \subset \mathcal{B}(F(m), r)$ for all $k_u \geq k_{u0}$.

Since $f^{k_u}(W)$ is always inside one chart neighbourhood, we do not need any scalings, and we may define $s_{m,k_u} = \mathrm{id}$ for all $k_u \geq k_{u0}$. Also we may assume the mappings ψ_{m,k_u} be of the form $\psi_{m,k_u} = t_{m,k_u} \circ \psi$. So we have a local representation

$$g_{m,k_u} = t_{m,k_u} \circ \psi \circ f^{k_u} \circ \varphi_m.$$

Since the $f^k \to F$, the translations $t_{m,k} : x \mapsto x - \psi(f^k(m))$ converge to $T : x \mapsto x - \psi F(m)$. Therefore (g_{m,k_u}) converges locally uniformly to $T \circ \psi \circ F \circ \varphi_m^{-1}$ in the neighbourhood $\varphi_m(W)$ of $\varphi(m)$.

□

Lemma 4.9. *The sequence $(g_{m,k})$ has in the set $\varphi_m(U) \subset \varphi_m(\tilde{U}_m)$ a subsequence converging uniformly to a constant if and only if the sequence (f^k) has a subsequence converging uniformly to a constant in U.*

Proof. Assume that the subsequence (f^{k_j}) of the iterates of f converges to a point p. Then after some index j_0, we can define the local representations in $\varphi_m(U)$ with $s_{m,k_j} = id$. The reason for this is that the scalings were needed to make sure that the image of the domain of definition stays inside one chart neighbourhood. Thus when $j \geq j_0$, we can define local representations as $g_{m,k_j} = \psi_{m,k_j} \circ f^{k_j} \circ \varphi_m^{-1}$. In fact since we stay in the same chart neighbourhood, $\psi_{m,k_j} = t_{m,k_j} \circ \psi$, meaning that the chart mappings differ only by translations. The translations do not change distance between points, and so when (f^{k_j}) converges uniformly to a point, also $g_{m,k_j} = t_{m,k_j} \circ \psi \circ f^{k_j} \circ \varphi_m^{-1}$ converges uniformly to a point in $\varphi_m(U)$, namely to the origin.

Assume then that a subsequence (g_{m,k_j}) of the sequence $(g_{m,k})$ converges uniformly to a constant. Since $g_{m,j}(m) = 0$, this point must in fact be the origin. In the proof of Lemma 4.8 we showed that since (g_{m,k_j}) converges, we can write its subsequence (g_{m,k_u}) as

$$g_{m,k_u} = s_m^{-1} \circ t_{m,k_u} \circ \psi \circ f^{k_u} \circ \varphi_m^{-1} \circ s_m.$$

Now we know especially that (g_{m,k_u}) converges uniformly to the origin. Therefore for all $\delta > 0$ there exists a u_0 such that when $u \geq u_0$, $g_{m,k_u}(\varphi_m(U)) \subset \mathbb{B}(0,\delta)$. The chart mappings are bilipschitz-continuous and the translations do not change distances between points, for any $\epsilon > 0$ we have $f^{k_u}(U) \subset \mathcal{B}(f^{k_u}(m),\epsilon)$, when k_u is great enough. Since (f^{k_u}) converges uniformly to a mapping F, in fact $f^{k_u}(U) \subset \mathcal{B}(F(m),\epsilon)$, which means that also (f^{k_u}) converges uniformly to a constant.

□

With the previous lemmas we can now prove the non-emptiness of the Julia sets of uniformly quasiregular mappings on riemannian manifolds. We define the degree of a uqr-mapping $f : M \to M$ as the greatest number of preimage points that any point on M has, that is, $\deg(f) = \sup_{p \in M} \#\{f^{-1}(p)\}$. (See the definition and results for "multiplicity" in [V1].)

Theorem 4.10. *Let $f : M \to M$ be a uniformly quasiregular mapping, with* $\deg(f) \geq 2$, *on a compact riemannian manifold M. Then the Julia set $\mathcal{J}_f$ of the mapping f is non-empty.*

Proof. Assume that the Julia set is empty. Then, by Definition 4.7 of the Julia sets, for any point $m \in M$ the sequence $(g_{m,k})$ is normal and thus contains a locally uniformly convergent subsequence $g_{m,k_j} \xrightarrow[k_j \to \infty]{} G_m$ in a neighbourhood $\varphi_m(U_m)$ of the origin. Since it is known that the limit of a locally uniformly convergent sequence of K-quasiregular mappings is either K-quasiregular or constant [R2, p. 157], the mapping G_m must be quasiregular or constant.

The manifold M can be covered with neighbourhoods U_i such that in each neighbourhood $f|U_i$ is non-injective and the limit mapping G_i exists: Consider a point $q \in M$. If f is a homeomorphism in an open set U_q, there exists another point $p \in M$ such that $p \in \{f^{-1}(f(q))\}$, since the set where $\deg(f) = 1$ has no interior points by the continuity of f. Similarly as we did on the page 25 for one intersection, we can redefine the local representations in a set U_{qp}, which consists of a finite chain of neighbourhoods joining the points p and q, and we see that a limit G_{qp} exists in the set U_{qp}.

Assume that U_m is one of these neighbourhoods U_i. Thus by Lemma 4.8, the sequence (f^k) has a locally uniformly converging subsequence (f^{k_j}) in U_m, that is, $f^{k_j} \xrightarrow[k_j \to \infty]{} F$ for some limit map F. Since $\deg(f) \geq 2$ and the degree grows when we iterate the mapping, we have $\deg(F) = \infty$. This follows from the facts that f is quasiregular, not a homeomorphism in U_m and the convergence is uniform. Thus there exists some point $y \in M$ such that the number of its preimages $\{F^{-1}(y)\}$ is infinite. Since the manifold M is compact, this infinite set of points has an accumulation point x by the Bolzano—Weierstrass theorem.

Let us consider the local representation of (f^k) at the point x, that is, the sequence $(g_{x,k})_{k=1}^{\infty}$, where

$$g_{x,k} := s_{x,k}^{-1} \circ \psi_{\alpha_{x,k}} \circ f^k \circ \varphi_x^{-1} \circ s_{x,k} : \varphi_x(\tilde{U}_x) \longrightarrow \psi_{\alpha_{x,k}}(U_{\alpha_{x,k}}).$$

Since the number of the elements in $\{F^{-1}(y)\}$ is infinite and x is an accumulation point of this set, there exists a point $u \in U_{G_x}$ (the set U_{G_x} is defined as in 4.3) which has an infinite set of preimage points $\{G_x^{-1}(u)\}$. On the other hand, as in the proof of Lemma 3.3 we see that the number of preimage points must be finite, if the mapping is quasiregular and non-constant. So G_x must be constant whenever defined. Respectively, the mapping F must then be constant in the set $\tilde{U}_x$ (Lemma 4.9).

Let $A \subset M$ be the largest possible domain containing the neighbourhood $\tilde{U}_x$ and having F constant on A. Let $a \in \partial A$ an accumulation point of A. Consider the limit mapping $G_a : \varphi_a(\tilde{U}_a) \to \psi_{a,k}(U_{\psi_{a,k}})$. Since the mapping F is constant on $A \cap \tilde{U}_a$, also G_a is constant on this domain (Lemma 4.9). As a limit of uniformly quasiregular mappings G_a is either quasiregular or constant in its entire area of definition. Thus G_a must be constant on $\varphi_a(\tilde{U}_a)$, too. This implies that F is also constant on $\tilde{U}_a$, which cannot be true: $\tilde{U}_a \cap \complement A \neq \emptyset$, but A was chosen so that F is constant only on A. Thus A must contain the whole area of definition of F.

Therefore, for any $r > 0$ we can choose indices k_i such that $f^{k_i}(U_i) \subset \mathcal{B}(c_i, r)$ for all i. We choose r so small that $\operatorname{diam}(f^j(\mathcal{B}(c_i, r))) < (\operatorname{diam} M)/lk_r$, where $l = \#U_i$, $j = 0, ..., k_r := k_+ - k_-$, $k_+ = \max\{k_1, ..., k_l\}$ and $k_- = \min\{k_1, ..., k_l\}$. Then we have $f^{k_+}(U_i) = f^{k_+ - k_i} f^{k_i}(U_i) \subset f^{k_+ - k_i}\mathcal{B}(c_i, r)$ for all i, $f^{k_+}(M) \subset f^{k_+}(\cup_i U_i) \subset \cup_{j=0...k_r} \cup_i f^j(\mathcal{B}(c_i, r))$, and finally

$$\operatorname{diam}(f^{k_+}(M)) \leq \operatorname{diam}(\cup_j \cup_i f^j(\mathcal{B}(c_i, r))) \leq \sum_j \sum_i \operatorname{diam}(f^j(\mathcal{B}(c_i, r))) < \operatorname{diam}(M),$$

which means that f^{k_+} cannot be surjective on M. Thus, the limit mapping G_x cannot be either a constant or a quasiregular mapping in the neighbourhood of the accumulation point x. Therefore, the assumption $\mathcal{J}_f = \emptyset$ is false, and the Julia set is non-empty.

□

5. Rescaling principle

In 1975, Zalcman in his article [Za] presented a result about complex functions. This result led to some simple proofs to lemmas considering Montel's normal families. Later Miniowitz generalized the results to cover the case of quasiregular mappings. She obtained necessary and sufficient terms to when a family of mappings is normal. The proof for the case of mappings between $\mathbb{R}^n$ and $\mathbb{S}^n$ is originally by Miniowitz [Min], and it is represented in detail in [Ka].

We will now consider a similar principle for quasiregular mappings $f : \mathbb{R}^n \to M^n$, where M^n is a smooth, oriented n-dimensional riemannian manifold. We shall call the result the rescaling principle, though in an euclidean space it is also known as Zalcman's lemma or the Bloch–Brody principle. The rescaling principle for quasiregular mappings has previously been considered by Hinkkanen, Martin and Mayer in [HMM] and Iwaniec and Martin in [IM1, pp. 484-485]. In chapter 2 of the article [BH], Bonk and Heinonen briefly discussed the generalization of this principle to riemannian manifolds.

5.1. **Criteria for normality.** To be able to prove a rescaling lemma for quasiregular mappings $f : \mathbb{R}^n \to M$, criteria for normality are needed. We get sufficient conditions from Ascoli's theorem. It is more interesting to ask what exactly are the necessary conditions. We formulate the criteria as follows:

Theorem 5.1. *Let $\mathcal{F}$ be a family of K-quasiregular mappings $f_\nu : \mathbb{B}(0, R) \to M^n$, where $n \geq 2$ and $R > 0$. Then $\mathcal{F}$ is a normal family if and only if for any compact subset X of the ball $\mathbb{B}(0, R)$ there exists a constant $\mathcal{C} > 0$ such that*

$$d(f(x), f(y)) \leq \mathcal{C}|x - y|^\alpha$$

for any $x, y \in X$ and $f \in \mathcal{F}$, when α, $0 < \alpha < 1$, is a constant depending on the dilatation K and the dimension n.

Note that generally the constant α depends also on the bilipschitz constant of the coordinate charts chosen for M, but since M is closed, we can choose this constant to be arbitrarily close to one.

Proof. Let us first show that the conditions are sufficient. By the assumption we have $d(f(x), f(y)) \leq \mathcal{C}|x-y|^\alpha$. Thus for any $\epsilon > 0$ we have $d(f(x), f(y)) \leq \epsilon$, as long as we choose x and y close each other. Therefore, the family $\mathcal{F}$ is equicontinuous and, by Ascoli's theorem, normal.

The necessity of the conditions needs much more consideration. Assume that $F = \{f_\nu\}$ is a normal family of K-quasiregular mappings $f_\nu : \mathbb{B}(0, R) \to M^n$, $n \geq 2$. Let X be an arbitrary compact subset of the ball $\mathbb{B}(0, R)$. Then there exists $0 < r < R$ such that $X \subset \bar{\mathbb{B}}(0, r) \subset \mathbb{B}(0, R)$ and we may in fact assume, that X is a closed ball about the origin. The set $f_\nu(X) \subset M$ is also compact as an image of a compact set on a continuous mapping (Figure 4)

In chapter 2 we showed that one can choose an atlas with L-bilipschitz-continuous coordinate mappings for any compact riemannian manifold. Let $\mathcal{A}_2$ and $\mathcal{A}_1$ be two

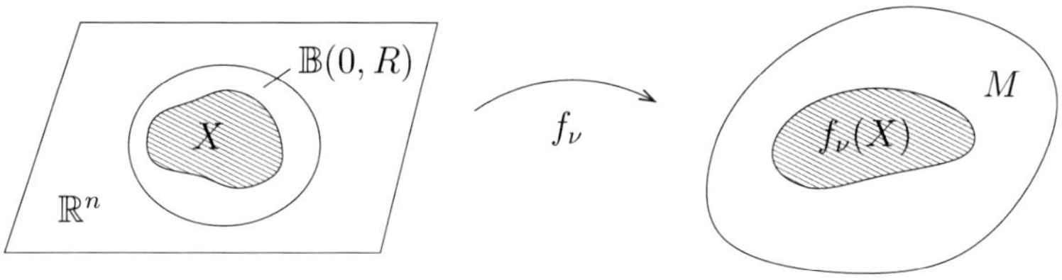

FIGURE 4. Compact subset X and its image $f_\nu(X)$.

such atlases as follows: Let

$$\mathcal{A}_2 = \{(\mathcal{B}(z_i, \frac{r_i}{2}), \varphi_i) \mid i \in I\}$$

be such a finite bilipschitz atlas for the manifold M that also

$$\mathcal{A}_1 = \{(\mathcal{B}(z_i, r_i), \varphi_i) \mid i \in I\}$$

is an atlas with the same indices and the same bilipschitz constant $L_{\mathcal{A}}$. The atlas $\mathcal{A}_2$ can be found since by the results from chapter 2, for any point z_i there exists a neighbourhood $\mathcal{B}(z_i, r_i)$ which can be mapped with a bilipschitz-continuous coordinate chart φ_j into $\mathbb{R}^n$. Consider the open cover $\{\mathcal{B}(z_i, \frac{r_i}{2})\}$. Since the manifold M is compact, there exists a finite subcover $M \subset \cup_{i \in I} \mathcal{B}(z_i, \frac{r_i}{2})$, where every neighbourhood is a coordinate neighbourhood. Thus we get a finite atlas $\mathcal{A}_2 = \{(\mathcal{B}(z_i, \frac{r_i}{2}), \varphi_i) \mid i \in I\}$, where the coordinate charts are restrictions of larger coordinate charts $(\mathcal{B}(z_i, r_i), \varphi_i)$. Since the atlases are finite, there exists a constant $\delta > 0$ such that $\varphi_i(\mathcal{B}(z_i, r_i)) \subset \mathbb{B}(0, \delta)$ for any $i \in I$.

Let U_X be such an open neighbourhood of the set X that $\bar{U}_X \subset \mathbb{B}(0, R)$. Then $\{f_\nu(U_X) \cap \mathcal{B}(z_i, \frac{r_i}{2}) \mid i \in I\}$ and $\{f_\nu(U_X) \cap \mathcal{B}(z_i, r_i) \mid i \in I\}$ are covers for the image set $f_\nu(U_X)$. Denote

$$V_i^1 := f_\nu(U_X) \cap \mathcal{B}(z_i, r_i),$$
$$V_i^2 := f_\nu(U_X) \cap \mathcal{B}(z_i, \frac{r_i}{2})$$

and

$$\tilde{V}_i^1 := f_\nu^{-1}(V_i^1) \text{ and } \tilde{V}_i^2 := f_\nu^{-1}(V_i^2).$$

Then $\{\tilde{V}_i^1 \mid i \in I\}$ are $\{\tilde{V}_i^2 \mid i \in I\}$ are covers for the set X, and it holds that $\tilde{V}_i^2 \subset \tilde{V}_i^1$ for any $i \in I$.

Consider then the set $\tilde{V}_j^1 = f_\nu^{-1}(V_j^1) \subset \mathbb{B}(0, R) \subset \mathbb{R}^n$, $j \in I$. In this set we can define a composite mapping

$$\tilde{f}_{\nu,j} := \varphi_j \circ f_\nu : \tilde{V}_j^1 \to \varphi_j(V_j^1) \subset \mathbb{R}^n,$$

where φ_j is the coordinate chart of the coordinate neighbourhood $\mathcal{B}(z_j, r_j)$ (Figure 5).

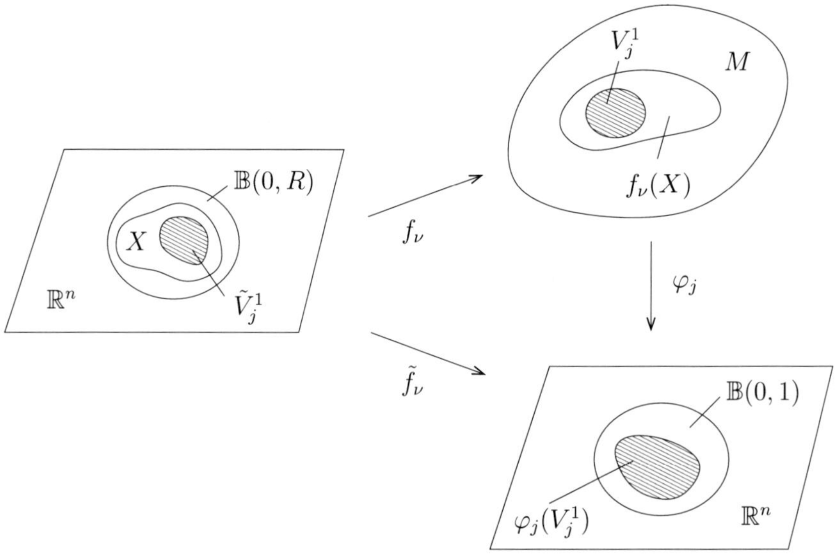

FIGURE 5. We restrict our considerations into a subset $\tilde{V}_j^1$ of the set $\mathbb{B}(0,R)$.

These mappings $\tilde{f}_{\nu,j}$ are $\tilde{K}$-quasiregular, where the constant $\tilde{K}$ depends on the dilatation of the mapping f_ν and the bilipschitz constant $L_{\mathcal{A}}$ of the atlas $\mathcal{A}_1$. Especially, it does not depend on the indices j and ν. In addition, we wish to point out that $\tilde{f}_{\nu,j}$ is bounded, since $\tilde{f}_{\nu,j}(\tilde{V}_j^1) \subset \mathbb{B}(0,\delta) \subset \mathbb{R}^n$.

Let G_j be an arbitrary compact subset of the set $\tilde{V}_j^1$. The mapping $\tilde{f}_{\nu,j}$ fulfils the demands of Theorem 3.2 of the article [MRV2]. Thus by this theorem,

$$d(\tilde{f}_{\nu,j}(x), \tilde{f}_{\nu,j}(y)) \leq \tilde{C}|x-y|^\alpha,$$

where $\tilde{C}$ is a constant, $\alpha = \tilde{K}^{\frac{1}{1-n}}$, $y \in \tilde{V}_j^1$, $x \in G_j$. Theorems 2.8 and 3.1 of the same article define the constant $\tilde{C}$: it depends on the mapping $\tilde{f}_{\nu,j}$, the dimension n, the dilatation $\tilde{K}$ and the sets G_j and $\tilde{V}_j^1$:

$$\tilde{C} = \lambda_{n,\tilde{K}} d(G_j, \partial\tilde{V}_j^1)^{-\alpha} d(\tilde{f}_{\nu,j}(\tilde{V}_j^1)).$$

We get a similar inequality also for the mapping f_ν. Since the coordinate mappings are bilipschitz-continuous with the constant $L_{\mathcal{A}}$, we have

$$d(\tilde{f}_{\nu,j}(x), \tilde{f}_{\nu,j}(y)) = d(\varphi_j \circ f_\nu(x), \varphi_j \circ f_\nu(y)) \geq \frac{1}{L_{\mathcal{A}}} d(f_\nu(x), f_\nu(y)).$$

Therefore

$$d(f_\nu(x), f_\nu(y)) \leq L_{\mathcal{A}} d(\tilde{f}_{\nu,j}(x), \tilde{f}_{\nu,j}(y)) \leq C|x-y|^\alpha,$$

when $y \in \tilde{V}_j^1$, $x \in G_j$ and $C = \tilde{C} L_{\mathcal{A}}$.

Since the compact set $G_j \subset \tilde{V}_j^1$ is arbitrary, we can choose $G_j = \overline{\tilde{V}_j^2}$, which is compact as a bounded and closed set in $\mathbb{R}^n$. Thus

$$d(f_\nu(x), f_\nu(y)) \leq C|x-y|^\alpha \tag{9}$$

holds for some constant C when $y \in \tilde{V}_j^1$ and $x \in \tilde{V}_j^2$ for some $j \in I$ and $f_\nu \in \mathcal{F}$.

We still wish to prove that the points x and y need not be chosen from sets with the same index j. Choose $x \in \tilde{V}_j^2$ and $y \in X \setminus \tilde{V}_j^1$. Denote the geodesic between the points x and y by γ_{xy}. Some subset $\mathcal{A}_{xy}$ of the cover $\{V_i^2 \mid i \in I\}$ is a cover for $f_\nu(\gamma_{xy})$. We wish to approximate the distance between the points $f_\nu(x)$ and $f_\nu(y)$ from above. We follow the path $f_\nu(\gamma_{xy})$ starting from the point $f_\nu(x)$. Denote the sets V_i^2 by V_{xy}^l, $l = 1, \ldots, L_{xy}$ in the order which we meet them. We pick points $(f_\nu(y_l))_{l=1}^{L_{xy}}$ along the path $f_\nu(\gamma_{xy})$ in such a way that each $f_\nu(y_l)$ belongs to at least two sets of the cover $\mathcal{A}_{xy}$ (see Figure 6). If the path cuts ∂V_{xy}^l in more than two points, we take the points near the first and the last cut, denote them by y_l^1 and y_l^2. Since we defined the sets V_{xy}^l as $V_{xy}^l := f_\nu(U_X) \cap \mathcal{B}(z_l, \frac{r_l}{2})$, the distance between the points y_l^1 and y_l^2 cannot be more than r_l. Therefore we can continue to follow the path from the point y_l^2 onwards, thus forgetting the extra loops. We will not enter the same set V_{xy}^l again.

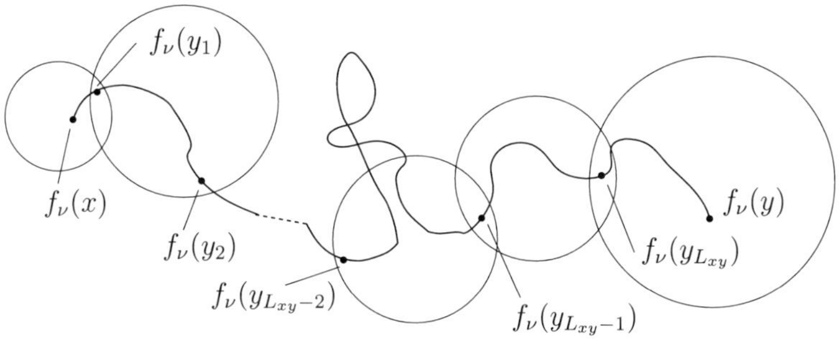

FIGURE 6. Points $f_\nu(y_l)$ chosen from the path $f_\nu(l_{xy})$.

We denote the preimages of the sets V_{xy}^l with the same indices: $f_\nu^{-1}(V_{xy}^l) := \tilde{V}_{xy}^l$. When we consider the preimages, we see that $x \in \tilde{V}_{xy}^1$, $y_1 \in \tilde{V}_{xy}^1 \cap \tilde{V}_{xy}^2$, $\ldots$, $y_l \in \tilde{V}_{xy}^l \cap \tilde{V}_{xy}^{l+1}$, $\ldots$, $y \in \tilde{V}_{xy}^{L_{xy}}$. Now we can use the inequality (9) in each set $\tilde{V}_{xy}^l$. We get

$$\begin{aligned} d(f_\nu(x), f_\nu(y)) &\leq d(f_\nu(x), f_\nu(y_1)) + d(f_\nu(y_1), f_\nu(y_2)) + \ldots + d(f_\nu(y_{L_{xy}}), f_\nu(y)) \\ &\leq C_0|x-y_1|^\alpha + C_1|y_1-y_2|^\alpha + \ldots + C_{L_{xy}}|y_{L_{xy}} - y|^\alpha \\ &\leq C_0|x-y|^\alpha + C_1|x-y|^\alpha + \ldots + C_{L_{xy}}|x-y|^\alpha \\ &= \max_{0 \leq k \leq L_{xy}} C_k(L_{xy}+1)|x-y|^\alpha. \end{aligned}$$

The last of the inequalities holds since $\alpha > 0$. Note that there exists a maximum for the constant L_{xy}, independent of the points x and y: for any pair x, y the value of $L_{xy} \leq L'$, where L' is the number of the elements in the finite cover $\{V_i^2 \mid i \in I\}$. Denote

$$\mathcal{C} := \max_{0 \leq k \leq L'} C_k(L'+1).$$

Since $X \subset \cup_{j \in I} \tilde{V}_j^2$ and $X \subset \cup_{j \in I} \tilde{V}_j^1$, we have obtained the inequality

$$d(f_\nu(x), f_\nu(y)) \leq \mathcal{C}|x-y|^\alpha$$

for any $x, y \in X$, $f_\nu \in \mathcal{F}$. □

5.2. **Rescaling principle on manifolds.**

Theorem 5.2 (Rescaling principle). *Let $\mathcal{F}$ be a family of K-quasiregular mappings $f : \Omega \to M^n$, where M^n is a closed compact riemannian manifold, and $\Omega \subset \mathbb{R}^n$ a domain. If the family $\mathcal{F}$ is not equicontinuous at a point $a \in \Omega$, there exists a sequence of real numbers $r_j \searrow 0$, a sequence of points $a_j \to a$, a sequence of mappings $\{f_j\} \subset \mathcal{F}$ and a non-constant K-quasiregular mapping $h : \mathbb{R}^n \to M^n$ such that*

$$f_j(r_j x + a_j) \to h(x)$$

locally uniformly in $\mathbb{R}^n$. Especially M^n is K-quasiregularly elliptic.

By a *K-quasiregularly elliptic* manifold we mean such an oriented riemannian n-manifold M that there exists a non-constant K-quasiregular mapping $f : \mathbb{R}^n \to M^n$.

A few parts of the following lengthy proof will be presented as lemmas to make the proof more understandable.

Proof. For simplicity, we assume $\Omega = \mathbb{B}(0,1)$ and the point a to be the origin: Let $\mathcal{F}$ be a family of K-quasiregular mappings $f : \mathbb{B}(0,1) \to M$ such that $\mathcal{F}$ is not equicontinuous at the origin. Let $(r_\nu)_{\nu=1}^\infty$ be a sequence of real numbers with $r_\nu \to 0$, when $\nu \to \infty$. Since the family $\mathcal{F}$ is not normal, by Theorem 5.1 for any $\nu \geq 1$ there exists a compact set $E_\nu \subset \mathbb{B}(0, r_\nu)$ such that for some $0 < \alpha < 1$

$$Q_f(x) := \sup_{|y| \leq 1} \frac{d(f(x), f(y))}{|x-y|^\alpha},$$

where $f \in \mathcal{F}$, is not bounded in E_ν.

Let us now define the elements of the sequence $(r_\nu)_{\nu=1}^\infty$ and also a sequence of points $(x_\nu)_{\nu=1}^\infty$. Set $r_1 = \frac{1}{2}$ and choose $x_1 \in \mathbb{B}(0, r_1)$ such that $Q_{f_{\mu_1}}(x_1) \geq 2$ for some $f_{\mu_1} \in \mathcal{F}$. We can always find such a point x_1, since $Q_{f_{\mu_1}}$ is unbounded in some compact set $E_1 \subset \mathbb{B}(0, r_1)$.

We define inductively the elements r_ν and x_ν, $\nu > 1$. Assume that $r_{\nu-1}$ and $x_{\nu-1} \in \mathbb{B}(0, r_{\nu-1})$ are defined. Then choose r_ν and x_ν as follows (see Figure 7):

1. If $|x_{\nu-1}| \geq \frac{1}{4} r_{\nu-1}$, choose r_ν such that

$$\frac{1}{4}|x_{\nu-1}| \leq r_\nu \leq \frac{1}{2}|x_{\nu-1}|$$

and $x_\nu \in \mathbb{B}(0, r_\nu)$ such that

(10)
$$Q_{f_{\mu_\nu}}(x_\nu) \geq \frac{2\nu^\alpha}{r_\nu^\alpha}$$

for some $f_{\mu_\nu} \in \mathcal{F}$.

2. If $|x_{\nu-1}| < \frac{1}{4} r_{\nu-1}$, choose $r_\nu = \frac{1}{4} r_{\nu-1}$ and $x_\nu = x_{\nu-1}$.

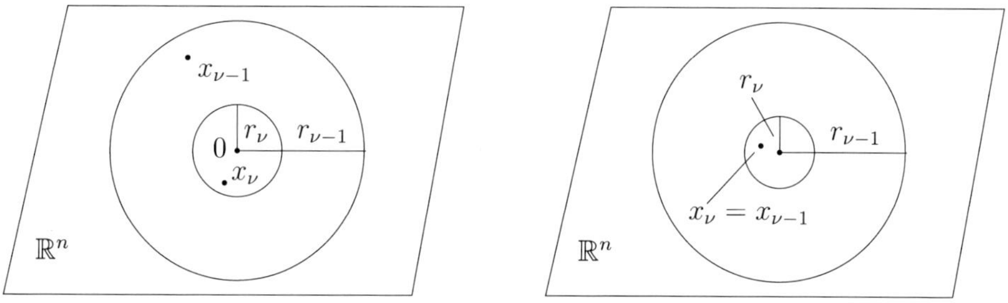

(a) Case 1: $|x_{\nu-1}| \geq \frac{1}{4} r_{\nu-1}$ (b) Case 2: $|x_{\nu-1}| < \frac{1}{4} r_{\nu-1}$

FIGURE 7. Choosing the radii r_ν and the points x_ν.

Obviously, $r_\nu \to 0$ when $\nu \to \infty$. Also the sequence (x_ν) converges to the origin: in the construction we may have to choose some of the points to be the same as the previous ones, but after some finite number N we always have $|x_\nu| \geq \frac{1}{4} r_{\nu+N} = (\frac{1}{4})^N r_\nu$, and so $x_{\nu+N+1}$ will be chosen inside the ball $\mathbb{B}(0, r_\nu)$, closer to the origin as $x_\nu = x_{\nu+1} = \ldots = x_{\nu+N}$.

Define

(11)
$$\Psi_\nu(x) := Q_{f_{\mu_\nu}}(x) \Big(\frac{\operatorname{dist}(x, \partial\mathbb{B}(x_\nu, r_{\nu+1}))}{r_{\nu+1}} \Big)^\alpha$$

and choose $a_\nu \in \mathbb{B}(x_\nu, r_{\nu+1})$ such that

(12)
$$\Psi_\nu(a_\nu) > \frac{1}{2} \sup_{x \in \mathbb{B}(x_\nu, r_{\nu+1})} \Psi_\nu(x).$$

Then define

(13)
$$R_\nu := \Psi_\nu(a_\nu)^{1/\alpha} r_{\nu+1}.$$

Now we can estimate the radii R_ν with the help of the inequalities (10) and (12) and the definition (11) as follows.

$$
\begin{aligned}
R_\nu &= \Psi_\nu(a_\nu)^{1/\alpha} r_{\nu+1} \\
&> (\frac{1}{2})^{1/\alpha}\Psi_\nu(x_\nu)^{1/\alpha} r_{\nu+1} = \frac{1}{2^{1/\alpha}} Q_{f_{\mu_\nu}}(x_\nu)^{1/\alpha} r_{\nu+1} \\
&\geq \frac{1}{2^{1/\alpha}}(\frac{2\nu_*^\alpha}{r_{\nu_*}^\alpha})^{1/\alpha} r_{\nu+1} = \nu_* \frac{r_{\nu+1}}{r_{\nu_*}} \\
&\geq \nu_* \frac{1}{16},
\end{aligned}
\tag{14}
$$

where ν_* is the largest such index $\nu_* < \nu+1$ for which x_{ν_*} and $x_{\nu+1}$ are two different points. The last inequality follows from the fact that either $r_{\nu+1} = \frac{1}{4} r_\nu$ (case 2) or we have

$$
\frac{r_{\nu+1}}{r_\nu} \geq \frac{r_{\nu+1}}{4|x_\nu|} \geq \frac{\frac{1}{4}|x_\nu|}{4|x_\nu|} = \frac{1}{16}
$$

(case 1). Thus $r_{\nu+1} > \frac{1}{16} r_{\nu_*}$, since $\nu_* < \nu+1$. When $\nu \to \infty$, also $\nu_* \to \infty$, and therefore

$$
R_\nu \to \infty.
$$

Define

$$
\rho_\nu := Q_{f_{\mu_\nu}}(a_\nu)^{-1/\alpha}. \tag{15}
$$

For any $|x| \leq R_\nu$ we have

$$
|a_\nu + \rho_\nu x - x_\nu| \leq |a_\nu - x_\nu| + \rho_\nu |x|,
$$

where by the definitions (15), (13) and (11) it holds that

$$
\begin{aligned}
\rho_\nu |x| &\leq Q_{f_{\mu_\nu}}(a_\nu)^{-1/\alpha} R_\nu \\
&= Q_{f_{\mu_\nu}}(a_\nu)^{-1/\alpha} \Psi_\nu(a_\nu)^{1/\alpha} r_{\nu+1} \\
&= \Big(\frac{\operatorname{dist}(a_\nu, \partial\mathbb{B}(x_\nu, r_{\nu+1}))}{r_{\nu+1}}\Big) r_{\nu+1} \\
&= \operatorname{dist}(a_\nu, \partial\mathbb{B}(x_\nu, r_{\nu+1})).
\end{aligned}
\tag{16}
$$

Therefore,

$$
|a_\nu + \rho_\nu x - x_\nu| \leq |a_\nu - x_\nu| + \operatorname{dist}(a_\nu, \partial\mathbb{B}(x_\nu, r_{\nu+1})) = r_{\nu+1}, \tag{17}
$$

since $a_\nu \in \mathbb{B}(x_\nu, r_{\nu+1})$. Thus the point $a_\nu + \rho_\nu x \in \mathbb{R}^n$ belongs to the ball $\mathbb{B}(x_\nu, r_{\nu+1})$ whenever $|x| \leq R_\nu$.

Respectively, for any $|y| \leq R_\nu$ we have

$$
\begin{aligned}
|a_\nu + \rho_\nu y| &\leq |a_\nu| + \rho_\nu |y| \\
&\leq |a_\nu| + Q_{f_{\mu_\nu}}(a_\nu)^{-1/\alpha} R_\nu \\
&= |a_\nu| + Q_{f_{\mu_\nu}}(a_\nu)^{-1/\alpha} \Psi_\nu(a_\nu)^{1/\alpha} r_{\nu+1} \\
&= |a_\nu| + \operatorname{dist}(a_\nu, \partial\mathbb{B}(x_\nu, r_{\nu+1})) \qquad (18)\\
&\leq |a_\nu| + r_{\nu+1} \leq |x_\nu| + 2r_{\nu+1} \\
&\leq r_\nu + 2r_{\nu+1} \leq 2r_\nu \\
&\leq 2r_1 = 1.
\end{aligned}
$$

In the above inequalities we have used (in addition to the definitions (15), (13) and (11)) the fact that a_ν belongs the ball $\mathbb{B}(x_\nu, r_{\nu+1})$. Thus the point $a_\nu + \rho_\nu y$ belongs to the unit ball , when $|y| \leq R_\nu$. By (18), we also see that the sequence of points $(a_\nu + \rho_\nu y)$ converges to the origin, since $r_\nu \to 0$.

Since $a_\nu + \rho_\nu x \in \mathbb{B}(x_\nu, r_{\nu+1})$, the mapping $x \mapsto f_{\mu_\nu}(a_\nu + \rho_\nu x)$ is defined for any $x \in \mathbb{B}(0, R_\nu)$. Therefore, we can define between an euclidean space and the manifold M the mappings

$$g_\nu : \bar{\mathbb{B}}(0, R_\nu) \to M^n, \; g_\nu(x) := f_{\mu_\nu}(a_\nu + \rho_\nu x),$$

where $\nu \in \mathbb{N}$ (see Figure 8).

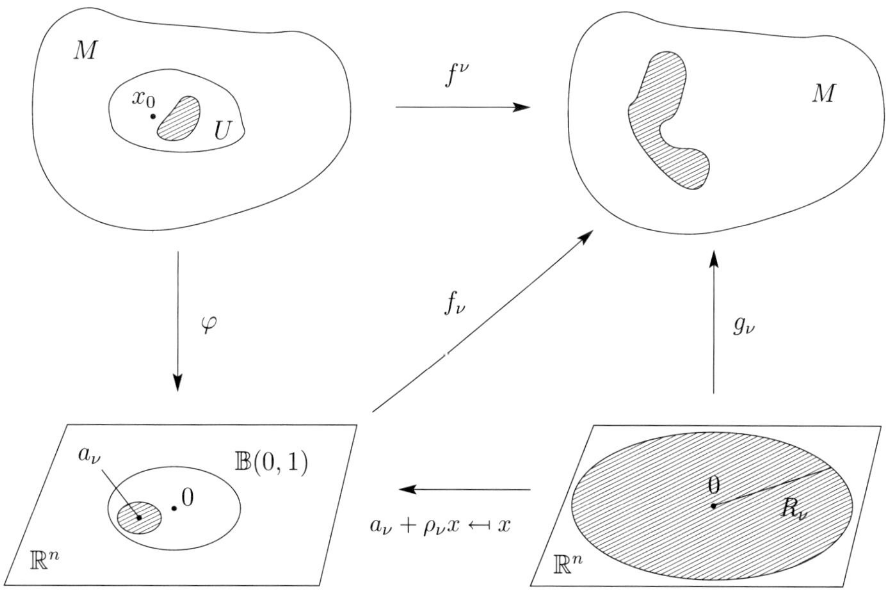

FIGURE 8. The mapping $g_\nu : \bar{\mathbb{B}}(0, R_\nu) \to M$.

Lemma 5.3. *When $x, y \in \mathbb{B}(0, \frac{R_\nu}{2})$, we have*

$$d(g_\nu(x), g_\nu(y)) \le C|x-y|^\alpha,$$

where ν is great enough, α is defined as earlier and C depends only on α.

Proof. Let us consider the variable $Q_{g_\nu}(x)$. Let $\nu > \nu_0$, where $|R_{\nu_0}| > 1$, and $x \in \mathbb{B}(0, R_\nu)$. Then it holds that

$$\begin{aligned} Q_{g_\nu}(x) &= \sup_{|y|\le 1} \frac{d(g_\nu(x), g_\nu(y))}{|x-y|^\alpha} \le \sup_{y\in\mathbb{B}(0,R_\nu)} \frac{d(g_\nu(x), g_\nu(y))}{|x-y|^\alpha} \\ &= \sup_{y\in\mathbb{B}(0,R_\nu)} \frac{d(f_{\mu_\nu}(a_\nu+\rho_\nu x), f_{\mu_\nu}(a_\nu+\rho_\nu y))}{(\frac{1}{\rho_\nu}|(a_\nu+\rho_\nu x)-(a_\nu+\rho_\nu y)|)^\alpha} \\ &\le Q_{f_{\mu_\nu}}(a_\nu+\rho_\nu x)\rho_\nu^\alpha, \end{aligned} \tag{19}$$

since for any $|y| < R_\nu$ we have $|a_\nu + \rho_\nu y| \le 1$ by Inequality (18). By the definitions (15) and (11) and the inequality (12) we get

$$\begin{aligned} Q_{g_\nu}(x) &\le \frac{Q_{f_{\mu_\nu}}(a_\nu+\rho_\nu x)}{Q_{f_{\mu_\nu}}(a_\nu)} \\ &= \frac{\Psi_\nu(a_\nu+\rho_\nu x)\Big(\frac{r_{\nu+1}}{\operatorname{dist}(a_\nu+\rho_\nu x, \partial\mathbb{B}(x_\nu, r_{\nu+1}))}\Big)^\alpha}{\Psi_\nu(a_\nu)\Big(\frac{r_{\nu+1}}{\operatorname{dist}(a_\nu, \partial\mathbb{B}(x_\nu, r_{\nu+1}))}\Big)^\alpha}. \end{aligned} \tag{20}$$

Because a_ν was chosen in such a way that $\Psi_\nu(a_\nu) > \frac{1}{2}\sup_{x\in\mathbb{B}(x_\nu,r_{\nu+1})}\Psi_\nu(x)$ (inequality (12)), we have

$$\Psi_\nu(a_\nu) > \frac{1}{2}\Psi_\nu(a_\nu+\rho_\nu x),$$

since $a_\nu + \rho_\nu x \in \mathbb{B}(x_\nu, r_{\nu+1})$. Thus it holds that

$$\begin{aligned} Q_{g_\nu}(x) &< 2\Big(\frac{\operatorname{dist}(a_\nu, \partial\mathbb{B}(x_\nu, r_{\nu+1}))}{\operatorname{dist}(a_\nu+\rho_\nu x, \partial\mathbb{B}(x_\nu, r_{\nu+1}))}\Big)^\alpha \\ &< 2\Big(\frac{\operatorname{dist}(a_\nu, \partial\mathbb{B}(x_\nu, r_{\nu+1}))}{\frac{1}{2}\operatorname{dist}(a_\nu, \partial\mathbb{B}(x_\nu, r_{\nu+1}))}\Big)^\alpha \\ &= 2^{1+\alpha}. \end{aligned} \tag{21}$$

The last inequality follows from the fact that

$$\rho_\nu|x| \le \frac{1}{2}\operatorname{dist}(a_\nu, \partial\mathbb{B}(x_\nu, r_{\nu+1})),$$

when $|x| \leq \frac{R_\nu}{2}$ (inequality (16) with radius $\frac{R_\nu}{2}$), and so

(22)
$$\begin{aligned}\operatorname{dist}(a_\nu + \rho_\nu x, \partial\mathbb{B}(x_\nu, r_{\nu+1})) &\geq \operatorname{dist}(a_\nu, \partial\mathbb{B}(x_\nu, r_{\nu+1})) - \rho_\nu|x| \\ &\geq \operatorname{dist}(a_\nu, \partial\mathbb{B}(x_\nu, r_{\nu+1})) - \frac{1}{2}\operatorname{dist}(a_\nu, \partial\mathbb{B}(x_\nu, r_{\nu+1})) \\ &= \frac{1}{2}\operatorname{dist}(a_\nu, \partial\mathbb{B}(x_\nu, r_{\nu+1})).\end{aligned}$$

We have thus shown that

$$Q_{g_\nu}(x) = \sup_{y \in \mathbb{B}(0,R_\nu)} \frac{d(g_\nu(x), g_\nu(y))}{|x-y|^\alpha} < 2^{1+\alpha},$$

when $x \in \mathbb{B}(0, \frac{R_\nu}{2})$. In other words,

$$d(g_\nu(x), g_\nu(y)) \leq 2^{1+\alpha}|x-y|^\alpha$$

for any $x, y \in \mathbb{B}(0, \frac{R_\nu}{2})$. □

In the sequence $(g_\nu)_{\nu=1}^\infty$ the next mapping is always defined in a bigger ball than the previous one. Let us consider some ball $\mathbb{B}(0, \frac{R_{\nu_0}}{2})$. In these balls all mappings g_ν, $\nu \geq \nu_0$, are defined, and by the previous lemma $d(g_\nu(x), g_\nu(y)) < C|x-y|^\alpha$ for any $x, y \in \mathbb{B}(0, \frac{R_{\nu_0}}{2})$. By Theorem 5.1 the family $(g_\nu)_{\nu=\nu_0}^\infty$ is normal in the ball $\mathbb{B}(0, \frac{R_{\nu_0}}{2})$.

We may assume that $(g_\nu)_{\nu=\nu_0}^\infty$ converges locally uniformly to a mapping h_{ν_0} in the ball $\mathbb{B}(0, \frac{R_{\nu_0}}{2})$ (take a subsequence if necessary). Respectively, the sequence $(g_\nu)_{\nu=\nu_0+1}^\infty$ converges locally uniformly to a mapping h_{ν_0+1} in $\mathbb{B}(0, \frac{R_{\nu_0+1}}{2})$. Since the limit mappings are unique, it holds that

$$h_{\nu_0+1}|_{\mathbb{B}(0,\frac{R_{\nu_0}}{2})} = h_{\nu_0}.$$

If we choose a larger starting index, we get a limit mapping which is defined in a bigger ball. At the limit when $\nu_0 \to \infty$ we get a mapping $g : \mathbb{R}^n \to M$ for which

$$g|_{\mathbb{B}(0,\frac{R_{\nu_0}}{2})} = \lim_{l\to\infty} g_{\nu_0+l}|_{\mathbb{B}(0,\frac{R_{\nu_0}}{2})}$$

for any ν_0.

We still have to prove that the limit mapping g is quasiregular and non-constant. To prove this we need the following lemma, which shows that the abnormality of the sequence (f_{μ_ν}) can especially be seen at the points a_ν.

Lemma 5.4. *$Q_{f_{\mu_\nu}}(a_\nu) \to \infty$, when $\nu \to \infty$.*

Proof. By the definition (11) and the inequality (12) we have

(23)

$$\begin{aligned}
Q_{f_{\mu_\nu}}(a_\nu) &= \Psi_\nu(a_\nu)\Big(\frac{r_{\nu+1}}{\operatorname{dist}(a_\nu, \partial\mathbb{B}(x_\nu, r_{\nu+1}))}\Big)^\alpha \\
&> \Big(\frac{r_{\nu+1}}{\operatorname{dist}(a_\nu, \partial\mathbb{B}(x_\nu, r_{\nu+1}))}\Big)^\alpha \frac{1}{2} \sup_{x\in\mathbb{B}(x_\nu, r_{\nu+1})} \Psi_\nu(x) \\
&= \frac{1}{2}\Big(\frac{r_{\nu+1}}{\operatorname{dist}(a_\nu, \partial\mathbb{B}(x_\nu, r_{\nu+1}))}\Big)^\alpha \sup_{x\in\mathbb{B}(x_\nu, r_{\nu+1})} \Big[Q_{f_{\mu_\nu}}(x)\Big(\frac{\operatorname{dist}(x, \partial\mathbb{B}(x_\nu, r_{\nu+1}))}{r_{\nu+1}}\Big)^\alpha\Big] \\
&= \frac{1}{2}\Big(\frac{1}{\operatorname{dist}(a_\nu, \partial\mathbb{B}(x_\nu, r_{\nu+1}))}\Big)^\alpha \sup_{x\in\mathbb{B}(x_\nu, r_{\nu+1})} \Big[Q_{f_{\mu_\nu}}(x)\operatorname{dist}(x, \partial\mathbb{B}(x_\nu, r_{\nu+1}))^\alpha\Big] \\
&\geq \frac{1}{2}\Big(\frac{1}{\operatorname{dist}(a_\nu, \partial\mathbb{B}(x_\nu, r_{\nu+1}))}\Big)^\alpha Q_{f_{\mu_\nu}}(x_\nu)\operatorname{dist}(x_\nu, \partial\mathbb{B}(x_\nu, r_{\nu+1}))^\alpha \\
&= \frac{1}{2}\Big(\frac{r_{\nu+1}}{\operatorname{dist}(a_\nu, \partial\mathbb{B}(x_\nu, r_{\nu+1}))}\Big)^\alpha Q_{f_{\mu_\nu}}(x_\nu),
\end{aligned}$$

and, in addition, since $a_\nu \in \mathbb{B}(x_\nu, r_{\nu+1})$, we have

$$r_{\nu+1} \geq \operatorname{dist}(a_\nu, \partial\mathbb{B}(x_\nu, r_{\nu+1})).$$

Therefore,

$$Q_{f_{\mu_\nu}}(a_\nu) \geq \frac{1}{2} Q_{f_{\mu_\nu}}(x_\nu),$$

and since the sequence (x_ν) was chosen in such a way that $Q_{f_{\mu_\nu}}(x_\nu) \to \infty$, also

$$Q_{f_{\mu_\nu}}(a_\nu) \to \infty$$

when $\nu \to \infty$. □

Corollary 5.5. *By Theorem 5.1 the sequence (f_{μ_ν}) is not normal.*

Theorem 5.6. *The limit mapping g is a non-constant quasiregular mapping.*

Proof. We divide the proof into three cases depending on whether the set $X = \{f_{\mu_\nu}(a_\nu) \in M \mid \nu \in \mathbb{N}\}$ is infinite, finite or just one point.

(1) Assume that $X = \{y_0\}$. We first prove that g is not constant and then that is must be quasiregular.

Assume that g is constant. By continuity we must then have $g(x) = y_0$ for any $x \in \mathbb{R}^n$, since $f_{\mu_\nu}(a_\nu) = y_0$ for any ν, and, on the other hand, $g_\nu(0) = f_{\mu_\nu}(a_\nu)$.

Now for any $\epsilon > 0$, $x \in \mathbb{B}(0, R_\nu)$ the point $g_\nu(x) = f_{\mu_\nu}(a_\nu + \rho_\nu x)$ belongs to the ball $\mathcal{B}(y_0, \epsilon)$ when the index ν is great enough.

By definition,

$$Q_{\mu_\nu}(a_\nu) = \sup_{|y|\leq 1} \frac{d(f_{\mu_\nu}(a_\nu), f_{\mu_\nu}(y))}{|a_\nu - y|^\alpha} = \sup_{|y|\leq 1} \frac{d(y_0, f_{\mu_\nu}(y))}{|a_\nu - y|^\alpha}.$$

Assume now that $y \notin \mathbb{B}(a_\nu, \rho_\nu R_\nu)$. Then $|y - a_\nu| \geq \rho_\nu R_\nu$, and we can write

$$\frac{d(f_{\mu_\nu}(a_\nu), f_{\mu_\nu}(y))}{|a_\nu - y|^\alpha} \leq \frac{d(f_{\mu_\nu}(a_\nu), f_{\mu_\nu}(y))}{\rho_\nu^\alpha R_\nu^\alpha} = Q_{\mu_\nu}(a_\nu) \frac{d(f_{\mu_\nu}(a_\nu), f_{\mu_\nu}(y))}{R_\nu^\alpha},$$

where we have used the definition (15).The distance between the points $f_{\mu_\nu}(a_\nu)$ and $f_{\mu_\nu}(y)$ is bounded by the size of the manifold. Thus for some ν',

$$d(f_{\mu_\nu}(a_\nu), f_{\mu_\nu}(y)) \leq \operatorname{diam} M \leq \frac{R_\nu^\alpha}{2}$$

when $\nu \geq \nu'$, since $R_\nu \to \infty$ monotonously. Now we have shown for all $y \notin \mathbb{B}(a_\nu, \rho_\nu R_\nu)$ that

$$\frac{d(f_{\mu_\nu}(a_\nu), f_{\mu_\nu}(y))}{|a_\nu - y|^\alpha} \leq \frac{Q_{\mu_\nu}(a_\nu)}{2},$$

when $\nu \geq \nu'$. Hence the supremum is obtained in the ball $\mathbb{B}(a_\nu, \rho_\nu R_\nu)$: for $\nu \geq \nu'$ it holds that

$$Q_{\mu_\nu}(a_\nu) = \sup_{y \in \mathbb{B}(a_\nu, \rho_\nu R_\nu)} \frac{d(f_{\mu_\nu}(a_\nu), f_{\mu_\nu}(y))}{|a_\nu - y|^\alpha}.$$

Now we fix a subsequence (ν_j) of $(\nu)_{\nu=\nu'}^\infty$ such that

$$Q_{\mu_{\nu_{j+1}}}(a_{\nu_{j+1}}) \geq 2Q_{\mu_{\nu_j}}(a_{\nu_j})$$

for any j. Suppose that $\varepsilon > 0$, and j_0 is such that $|a_{\nu_{j+k}} - a_{\nu_j}| < \varepsilon$ for any $j > j_0$, $k \geq 1$. Denote $y = a_{\nu_{j+k}} + \rho_{\nu_{j+k}} x \in \mathbb{B}(a_{\nu_{j+k}}, \rho_{\nu_{j+k}} R_0)$, $R_0 \in \mathbb{R}$, where we may assume (by restricting our considerations again to a subsequence, if necessary) that $d(y_0, f_{\mu_{\nu_{j+k}}}(y)) \leq d(y_0, f_{\mu_{\nu_j}}(y))$ for any y and any $k \geq 1$, since g_ν converges to the constant y_0.

It holds that

$$\begin{aligned} \frac{d(y_0, f_{\mu_{\nu_{j+k}}}(y))}{\varepsilon + |a_{\nu_{j+k}} - y|^\alpha} &\leq \frac{d(y_0, f_{\mu_{\nu_j}}(y))}{|a_{\nu_j} - y|^\alpha} \leq Q_{\mu_{\nu_j}}(a_{\nu_j}) \\ &\leq \frac{Q_{\mu_{\nu_{j+1}}}(a_{\nu_{j+1}})}{2} \leq \frac{Q_{\mu_{\nu_{j+k}}}(a_{\nu_{j+k}})}{2}. \end{aligned}$$

We can choose ε arbitrarily small. Thus,

(24) $$\frac{d(y_0, f_{\mu_{\nu_{j+k}}}(y))}{|a_{\nu_{j+k}} - y|^\alpha} \leq \frac{Q_{\mu_{\nu_{j+k}}}(a_{\nu_{j+k}})}{2}$$

for every $y \in \mathbb{B}(a_{\nu_{j+k}}, \rho_{\nu_{j+k}} R_0)$. Hence the supremun is not obtained when $|x| \leq R_0$ with any fixed R_0.

We may thus assume that $|x_\nu| \geq 1$ for all the points x_ν that satisfy

$$\frac{d(g_\nu(0), g_\nu(x_\nu))}{|\rho_\nu x_\nu|^\alpha} = \frac{d(f_{\mu\nu}(a_\nu), f_{\mu n u}(y))}{|a_\nu - y|^\alpha} \geq \frac{1}{2} Q_{\mu_\nu}(a_\nu),$$

where $y = a_\nu + \rho_\nu x_\nu$. Therefore

$$d(g_\nu(0), g_\nu(x_\nu)) \geq \frac{1}{2} Q_{\mu_\nu}(a_\nu)|\rho_\nu x_\nu|^\alpha = \frac{1}{2}|x_\nu|^\alpha \geq \frac{1}{2},$$

for any ν great enough. Thus the limit mapping g cannot be a constant.

We use Rickman's theorem [R2, p. 157] to prove that the limit mapping is quasiregular. The theorem is valid for mappings in an euclidean space, so we must add a coordinate chart to the mapping $g : \mathbb{R}^n \to M$. Since quasiregularity is a local property, it is sufficient to consider the situation near some arbitrary point $x \in \mathbb{R}^n$. Now $g_\nu(x) \to g(x)$, and we can choose such a neighbourhood U for the point $g(x)$ that U belongs to just one coordinate neighbourhood and that for any $\nu \geq \nu_1$ we have $g_\nu(U_x) \subset U$ for some neighbourhood $U_x \subset \mathbb{R}^n$ of the point x.

Define

$$h_\nu := \varphi \circ g_\nu : U_x \to \mathbb{R}^n,$$

where $\nu \geq \nu_1$ and φ is a bilipschitz-continuous coordinate mapping for the coordinate neighbourhood U. The mappings h_ν converge locally uniformly to a mapping $h := \varphi \circ g$ wherever defined. By Rickman's theorem [R2, p. 157] the mapping h is constant or quasiregular. Since we already proved that g cannot be constant, neither can h be. So h must be quasiregular, and thus also $g = \varphi^{-1} \circ h$ is quasiregular as a composition of a bilipschitz-continuous mapping and a quasiregular mapping.

(2) Assume that $X = \{f_{\mu_\nu}(a_\nu) \in M \mid \nu \in \mathbb{N}\}$ is finite. Now we get some point y_1 for infinitely many indices ν. Thus we can take such a subsequence (f_{μ_l}) of the sequence (f_{μ_ν}) that

$$f_{\mu_l}(0) = y_1$$

for any μ_l. In the proof of Lemma 5.4 we could as well have considered $Q_{f_{\mu_l}}(a_l)$ in the place of $Q_{f_{\mu_\nu}}(a_\nu)$ and noticed that also $Q_{f_{\mu_l}}(a_l) \to \infty$, when $l \to \infty$. Thus the sequence (f_{μ_l}) is not normal, and by taking a subsequence, we have again obtained the same situation as in the first case, where we assumed X to be just one point. Therefore, the limit mapping g, which is a limit for the subsequence (g_l) as well, is quasiregular and non-constant also when X is finite.

(3) Assume that the set X is infinite. Then by the Bolzano–Weierstrass theorem, the set X has an accumulation point y_2, since X is an infinite set on a compact manifold. In any neighbourhood of the point there exists an infinite number of points that belong to X. Let $\epsilon > 0$ be such that the ball $\mathbb{B}(y_2, \epsilon)$ belongs to some coordinate neighbourhood. Take such a subsequence (f_{μ_l}) from the sequence (f_{μ_ν}) that

$$f_{\mu_l}(0) \in \mathbb{B}(y_2, \frac{\epsilon}{2^l})$$

for any μ_l. As in the case 2, the claim of Lemma 5.4 holds for this subsequence, and (f_{μ_l}) is not normal in the neighbourhood of the origin.

Assume that the limit mapping g is constant. Then we must have $g(x) = y_2$ for any x, since g is continuous. For any $\mathbb{B}(y_2, \epsilon)$, there exists an index l_0 such that

$$f_{\mu_l}(a_l + \rho_l x) = g_l(x) \in \mathbb{B}(y_2, \epsilon),$$

whenever $l > l_0$. As in the case 1, this is in contradiction to the fact that (f_{μ_l}) is not normal. Thus the limit mapping is not normal.

The quasiregularity of the mapping g follows as in the case 1, since as a locally uniform limit of the sequence (g_ν) the mapping g is also the limit mapping for (g_l).

□

We have now proved the rescaling principle 5.2. □

5.3. **Manifolds supporting uqr-mappings are elliptic.** Now we are able to use the rescaling principle to prove our main theorem. Remember that by a quasiregularly elliptic manifold we mean such a manifold that there exists a quasiregular mapping from an euclidean space to this manifold.

Theorem 5.7. *Let M^n be a smooth, oriented and compact n-dimensional riemannian manifold and $f : M \to M$ a non-constant uniformly quasiregular mapping with non-empty branch set. Then there exists a quasiregular mapping $g : \mathbb{R}^n \to M$. In other words, M is quasiregularly elliptic.*

Proof. We wish to use the rescaling principle to prove the existence of the mapping g. So we need to construct a suitable family of quasiregular mappings to use the principle.

Let $f : M^n \to M^n$ be a K'-quasiregular mapping on a smooth, oriented, compact riemannian n-manifold M, with a non-empty branch set. By Theorem 4.10 we know that the Julia set $\mathcal{J}_f$ is non-empty. Therefore we can choose a point $x_0 \in \mathcal{J}_f$. Let $\varphi : U \to \mathbb{R}^n$ be such a bilipschitz-continuous coordinate mapping in some neighbourhood U of the point x_0 that $\varphi(x_0) = 0$ and $\varphi(U) = \mathbb{B}(0, 1)$ (see Figure 9).

Define a composite mapping

$$f_\nu := f^\nu \circ \varphi^{-1}|_{\mathbb{B}(0,1)} : \mathbb{B}(0, 1) \to M^n$$

from the iterates f^ν of f and the coordinate mapping φ. All the mappings f_ν, $\nu \in \mathbb{N}$, are K-quasiregular with the same constant $K = K(K', L)$, where L is the bilipschitz constant of the coordinate mapping φ and K' is the dilatation of the mapping f. The family of mappings $\mathcal{F} = \{f_\nu \mid \nu \in \mathbb{N}\}$ is not normal, since $x_0 = \varphi^{-1}(0) \in \mathcal{J}_f$, meaning that $0 \in \mathcal{J}_\mathcal{F}$.

From a uniformly quasiregular mapping $f : M \to M$ we have thus constructed a family $\mathcal{F} = \{f_\nu \mid \nu \in \mathbb{N}\}$ of K-quasiregular mappings $f_\nu = f^\nu \circ \varphi^{-1}|_{\mathbb{B}(0,1)} : \mathbb{B}(0, 1) \to M$, and the family $\mathcal{F}$ is not normal at the origin. Now we can use the rescaling principle 5.2, and as a limit mapping we get a K-quasiregular mapping $g : \mathbb{R}^n \to M$. □

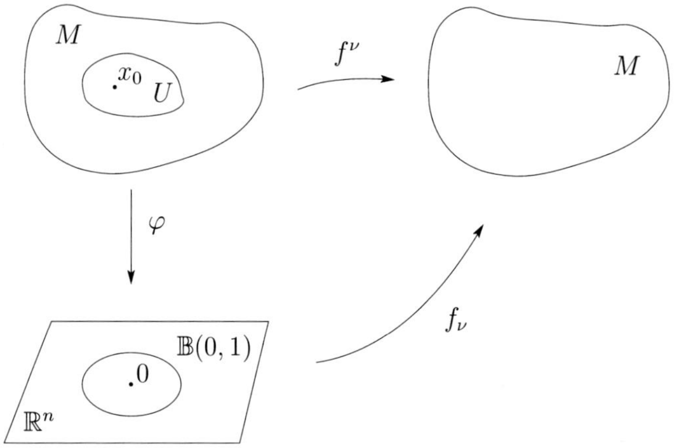

FIGURE 9. Mappings between the manifold M and the euclidean space $\mathbb{R}^n$.

6. Elliptic 3-manifolds

We have already proved that if there exists a non-trivial uniformly quasiregular mapping $f : M^n \to M^n$ on a riemannian n-manifold M^n, there also exists a quasiregular mapping $g : \mathbb{R}^n \to M^n$. In the following two chapters we will discuss a converse version of this theorem in three dimensions. We will prove that in three dimensions every quasiregularly elliptic manifold has a uniformly quasiregular mapping.

First we will recall how to represent a 3-manifold by a polyhedron. Then we will consider Thurston's geometrization conjecture and eight different 3-dimensional model geometries classified by W. Thurston [Th]. By a result of J. Jormakka [Jo], we will see that only three of those eight correspond to elliptic manifolds. Then what we need to do is to find a uniformly quasiregular mapping on each manifold modelled by these three model geometries.

6.1. **Representation by a polyhedron.** In 1895, Poincaré introduced a method of constructing 3-manifolds by identifying faces of simply connected polyhedra. Just as one can represent a surface by an appropriate polygon having pairwise associated sides, one can represent a class of 3-dimensional manifolds by 3-dimensional solid polyhedrons having pairwise associated surface faces.

The strict 3-dimensional analogy with a polygon schema in 2 dimensions is

(1) a finite set of polyhedra (topological 3-balls) called *cells*, with disjoint interiors,
(2) faces of cells identified in pairs, with vertices corresponding to vertices,
(3) resulting in a connected complex.

These three conditions do not guarantee that the outcome will be a manifold. It is called a 3-dimensional *pseudomanifold*, and it is a manifold if and only if it satisfies the following additional condition [St, Ch 8.2.1].

4. The neighbourhood surface of each vertex is a 2-sphere.

Let V, E, F and C denote the numbers of vertices, edges, faces and cells in a cell decomposition of a 3-dimensional pseudomanifold. Then $\chi(M) = V - E + F - C$ is called the *Euler characteristic* of M.

Theorem 6.1. *The Euler characteristic $\chi(M)$ of a 3-dimensional pseudomanifold M is zero if and only if M is a manifold.*

A quite elementary proof to this theorem is presented in [St, pp. 249-250]. The fact that 3-manifolds have the Euler characteristic 0 can be derived also from Poincaré's method of dual cell decomposition. The method in particular shows that any manifold of odd dimension has the Euler characteristic 0.

6.2. **3-dimensional model geometries.** Thurston's conjecture states that after a 3-manifold is splitten into its connected sum and the Jaco-Shalen-Johannson torus decomposition, the remaining components each admit exactly one of the 3-dimensional model geometries. We will now define these model geometries as well as the decompositions.

In dimension 3 we have only eight different model geometries. This classification was given by W. Thurston [Th] and has been discussed in great detail also by P. Scott [Sco]. But what do we mean by the term geometry? There are distinct but related approaches which one can take. For instance, we could think of a geometry as a space equipped with such notions as lines and planes, or with either metric or a riemannian metric, or with a notion of congruence. So it is obvious that we first need to define a geometry.

Definition 6.2. A *geometry* is a pair (X, G) where X is a manifold and G is a group of homeomorphisms called the isometry group of X which acts transitively on X with compact point stabilizers.

Recall that if a group G acts on a space X, the stabilizer of a point $x \in X$ is the subgroup G_x of G that leaves x invariant. Two geometries (X, G) and (X', G') are equivalent if there exists a diffeomorphism of X to X' conjugating the action of G onto the action of G'. In particular, G and G' must be isomorphic.

Let us now define a model geometry as in [Th, Def 3.8.1].

Definition 6.3. A *model geometry* (X, G) is a manifold X together with a Lie group G of diffeomorphisms of X, such that

a) X is connected and simply connected;
b) G acts transitively on X, with compact point stabilizers;
c) G is not contained in any larger group of diffeomorphisms of X with compact stabilizers of points; and
d) there exists at least one compact manifold modelled on X, G.

In two dimensions we have only three model geometries: spherical, euclidean and hyperbolic [Th, Thm 3.8.2]. Every 3-dimensional geometry can be modelled by one of the following eight model geometries.

Theorem 6.4 (3-dimensional model geometries). *There are eight different three-dimensional model geometries (X, G), as follows:*

a) If the point stabilizers are 3-dimensional, X is $\mathbb{S}^3$ (spherical), $\mathbb{R}^3$ (euclidean) or $\mathbb{H}^3$ (hyperbolic).

b) If the point stabilizers are 1-dimensional, then X fibres over one of the 2-dimensional model geometries in a way that is invariant under G. There is a riemannian metric on X such that the connection orthogonal to the fibres has curvature 0 or 1.

b_1) If the curvature is zero, X is $\mathbb{S}^2 \times \mathbb{R}$ or $\mathbb{H}^2 \times \mathbb{R}$.

b_2) If the curvature is 1, we have nilgeometry Nil which fibres over $\mathbb{R}^2$ or the geometry of $\widetilde{\mathcal{SL}}_2(\mathbb{R})$ which fibres over $\mathbb{H}^2$.

c) The only geometry with 0-dimensional stabilizers is solvegeometry Sol, which fibres over the line.

The proof of this theorem is presented by W. Thurston in [Th, pp. 182-188]. By this theorem Thurston showed that, in dimension three, there are eight possible model geometries, all of which are realized. Let us now summarize some basic features of these geometries as in [An]:

For the constant curvature geometries $\mathbb{H}^3$, $\mathbb{R}^3$ and $\mathbb{S}^3$, the space X is the simply connected space form $\mathbb{H}^3$, of constant curvature -1, or $\mathbb{R}^3$ of curvature 0, or $\mathbb{S}^3$ of curvature $+1$. The corresponding geometries are called hyperbolic, euclidean, and spherical. The groups G are $\mathrm{PSL}(2,\mathbb{C})$, $\mathbb{R}^3 \times \mathrm{SO}(3)$ and $\mathrm{SO}(4)$ with stabilizer $H_0 = \mathrm{SO}(3)$.

For the product geometries $X = \mathbb{S}^2 \times \mathbb{R}$ or $\mathbb{H}^2 \times \mathbb{R}$, the groups G are given by the subgroups of orientation preserving diffeomorphisms $\mathrm{Isom}\mathbb{H}^2 \times \mathrm{Isom}\mathbb{R}$ and $\mathrm{SO}(3) \times \mathrm{Isom}\mathbb{R}$ with stabilizer $H_0 = \mathrm{SO}(2)$.

The last three model geometries are the twisted products $\widetilde{\mathcal{SL}}_2(\mathbb{R})$, Nil, Sol. For $\widetilde{\mathcal{SL}}_2(\mathbb{R})$ the space X is the universal cover of the unit sphere bundle of $\mathbb{H}^2$, and $G = \widetilde{\mathcal{SL}}_2(\mathbb{R}) \times \mathbb{R}$ with $H_0 = \mathrm{SO}(2)$. For the nilgeometry, X is the 3-dimensional Heisenberg group, and G is the semidirect product of X with $\mathbb{S}$, acting by rotations on the quotient of X by its centre, $H_0 = \mathrm{SO}(2)$. For the solgeometry, X is the 3-dimensional solvable Lie group, $H_0 = \{e\}$, and G is an extension of X by an automorphism group of order eight.

Let us define the following decompositions needed for the geometrization conjecture. Detailed definitions of the concepts as well as proofs of these decompositions are presented in [An]. We recall that a closed 3-manifold is *prime* if it is not the three-sphere and cannot be written as a nontrivial connected sum of closed 3-manifolds. The first from the two following theorems is originally due to H. Kneser [Kn] and J. Milnor [Mil], the latter to W. Jaco and P. Shalen [JS] and K. Johannson [Jo].

Theorem 6.5 (Connected sum decomposition). *Let M be a closed, oriented 3-manifold. Then M has a finite decomposition as a connected sum*

$$M = M_1 \# M_2 \# \cdots \# M_k,$$

where each M_i is prime. The collection $\{M_i\}$ is unique up to the permutation of the factors.

Theorem 6.6 (Jaco-Shalen-Johannson torus decomposition). *Let M be a closed, oriented, irreducible 3-manifold. Then there is a finite collection of disjoint incompressible tori $T_i^2 \subset M$ that separate M into a finite collection of compact 3-manifolds with toral boundary, each of which is either torus-irreducible or Seifert-fibred. A minimal such collection (with respect to cardinality) is unique up to isotopy.*

Thus the connected sum decomposition means that every compact 3-manifold is the connected sum of a unique collection of prime 3-manifolds. The Jaco-Shalen-Johannson torus decomposition states that irreducible orientable compact 3-manifolds have a canonical (up to isotopy) minimal collection of disjointly embedded incompressible tori such that each component of the 3-manifold removed by the tori is either atoroidal or Seifert-fibred.

We are now able to state Thurston's geometrization conjecture. It proposes a complete characterization of geometric structures on three-dimensional manifolds.

Theorem 6.7 (Thurston's conjecture). *After a three-manifold is splitten into its connected sum and the Jaco-Shalen-Johannson torus decomposition, the remaining components each admit exactly one of the following model geometries*

- $\mathbb{S}^3$ *(spherical geometry)*
- $\mathbb{R}^3$ *(euclidean geometry)*
- $\mathbb{H}^3$ *(hyperbolic geometry)*
- $\mathbb{S}^2 \times \mathbb{R}$
- $\mathbb{H}^2 \times \mathbb{R}$
- $\widetilde{\mathcal{SL}}_2(\mathbb{R})$
- *Nil (nilgeometry)*
- *Sol (solgeometry)*

Results due to G. Perelman (2002, 2003) appear to establish Thurston's geometrization conjecture and thus also the Poincaré conjecture.

Now we have the eight model geometries for 3-dimensional manifolds. Next we wish to know which of these need to be considered. Fortunately, this question has already been solved: according to J. Jormakka, there are only three possible model geometries for an elliptic 3-manifold [Jo, Th 1.3].

Theorem 6.8. *If the riemannian covering space $\tilde{M}$ of a closed orientable riemannian 3-manifold M is one of the eight 3-dimensional simply connected homogeneous spaces $\mathbb{S}^3$, $\mathbb{S}^2 \times \mathbb{R}$, $\mathbb{R}^3$, Nil, Sol, $\widetilde{\mathcal{SL}}_2(\mathbb{R})$, $\mathbb{H}^2 \times \mathbb{R}$, or $\mathbb{H}^3$, there exists a nonconstant quasiregular mapping from the euclidean 3-space $\mathbb{R}^3$ to M if and only if $\tilde{M} = \mathbb{S}^3$, $\mathbb{S}^2 \times \mathbb{R}$ or $\mathbb{R}^3$.*

Jormakka makes a remark that the existence of a nonconstant quasiregular mapping f from $\mathbb{R}^n$ to an orientable riemannian n-manifold M^n or to an orientable riemannian manifold M_1^n which covers M^n are equivalent properties [Jo, p. 21].

7. UNIFORMLY QUASIREGULAR MAPPINGS ON ELLIPTIC 3-MANIFOLDS

At the end of the preceeding chapter we saw that the only closed elliptic 3-manifolds are those manifolds which are covered by $\mathbb{S}^3$, $\mathbb{S}^2 \times \mathbb{R}$ or $\mathbb{R}^3$. We shall now prove the following theorem by considering these three cases independently.

Theorem 7.1. *All quasiregularly elliptic compact oriented 3-dimensional riemannian manifolds admit a nontrivial uniformly quasiregular mapping.*

First we briefly state the known results for the case $\mathbb{S}^3$. The two other cases are the more interesting ones. For $\mathbb{R}^3$ we have to consider six different classes of manifolds and two for $\mathbb{S}^2 \times \mathbb{R}$. We will construct uniformly quasiconformal mappings of Lattè's type on these manifolds.

7.1. Spherical space forms.

We shall begin to prove Theorem 7.1 by considering the case $\mathbb{S}^3$. Let us first define spherical space forms (see [Wo] for complete classification).

Definition 7.2. A *spherical space form* M^n is a smooth, complete, connected riemannian manifold of constant sectional curvature $K > 0$.

Spherical space forms can be characterized as spaces isometric to quotients $\mathbb{S}^3/\Gamma$, where $\mathbb{S}^n = \mathbb{S}^n(\frac{1}{\sqrt{K}}) \subset \mathbb{R}^{n+1}$ is equipped with the standard metric and $\Gamma \subset O(n+1)$ acts freely and properly discontinuously. Examples of such spaces are the Poincaré homology sphere and 3-dimensional lens spaces $L(p,q)$. For more examples, see [Mo].

K. Peltonen proved the following two theorems in [Pe]. The first one is for spherical space forms, but it turns out that the metric is not an obstruction here. The situation is even more general.

Theorem 7.3. *Let M^n be a spherical space form. Then M^n admits a branched uniformly quasiregular mapping.*

Theorem 7.4. *Let M^n be a smooth riemannian manifold with universal covering space $\mathbb{S}^n$. Then M^n admits a branched uniformly quasiregular mapping.*

These theorems are proved by using a generalization of the so-called conformal trap method, due to G. Martin and T. Iwaniec (see [IM1]). The theorems hold in all the dimensions n, so also in three dimensions. Thus, by the conformal trap method, we can construct a nontrivial uniformly quasiregular mapping $f : M \to M$. The Julia sets of these uqr-mappings are Cantor sets [IM1, p. 500].

7.2. Euclidean space forms.

Let M be a 3-dimensional compact riemannian manifold which has $\mathbb{R}^3$ as a universal covering. Thus M is an *euclidean space form.* Such manifolds are called *flat*, since both their curvature and torsion are identically zero.

J. Wolf has considered the 3-dimensional euclidean space forms in his book [Wo]. We need the following result [Wo, Th 3.5.5]: There are just 6 affine diffeomorphism classes of compact connected flat 3-dimensional riemannian manifolds. They are represented by the manifolds $\mathbb{R}^3/\Gamma$, where Γ is one of the 6 groups G_i given below. Here $\{a_1, a_2, a_3\}$ is a set of linearly independent generators of the translation lattice, $t_i = t_{a_i} : x \mapsto x + a_i$ are the translations, and $\Psi = \Gamma/\Gamma^*$ is the linear holonomy group of the manifold. The subgroup $\Gamma^* = \Gamma \cap \mathbb{R}^3$ consisting of translations is a normal free abelian subgroup of rank 3 which is maximal abelian in Γ (see [Wo, p.115] for detailed definitions).

The translations $\{t_1, t_2, t_3\}$ are denoted by dash lines, solid lines and broken lines respectively in the following pictures. We now discuss these 6 manifolds in detail and construct a uniformly quasiregular mapping of Lattès type on each of them. We first consider the cases G_1, G_2, G_4 and G_6 where the underlaying plane lattice in $\mathbb{R}^3/\Gamma$ consists of tetragons, see Figure 10(a). Then we discuss G_3 and G_5 where the plane lattice is hexagonal as in Figure 10(b).

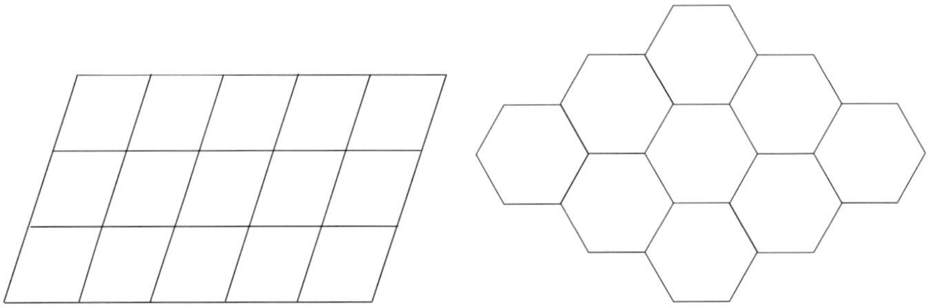

(a) Tetragonal lattice (b) Hexagonal lattice

FIGURE 10. Underlaying plane lattices for the cases $G_1 - G_6$.

7.2.1. *Manifolds with tetragonal basis.* G_1. $\Psi = \{1\}$ and Γ is generated by the translations $\{t_1, t_2, t_3\}$ with $\{a_i\}$ linearly independent (see Figure 11). This structure gives a polyhedron schema: just one cell, the cube, where we identify opposite faces. This manifold, let us denote it by M_1, is a three-dimensional torus.

G_2. $\Psi = \mathbb{Z}_2$ and Γ is generated by $\{\alpha, t_1, t_2, t_3\}$, where

$$\begin{cases} \alpha^2 = t_1 \\ \alpha t_2 \alpha^{-1} = t_2^{-1} \\ \alpha t_3 \alpha^{-1} = t_3^{-1} \end{cases}.$$

The generator a_1 is orthogonal to a_2 and a_3 while $\alpha = A \circ t_{a_1/2}$ with a linear mapping A defined by

$$\begin{cases} A(a_1) = a_1 \\ A(a_2) = -a_2 \\ A(a_3) = -a_3 \end{cases}.$$

Thus α consists of a translation $t_{a_1/2}$ and a rotation defined by A. In this polyhedron schema we have a cube where we identify opposite vertical faces and glue the top to the bottom of the cube with a twist of angle π. Again we have a manifold M_2, since the Euler characteristic is $\chi(M_2) = 1-3+3-1 = 0$. Just one block of Figure 11 covers our manifold once.

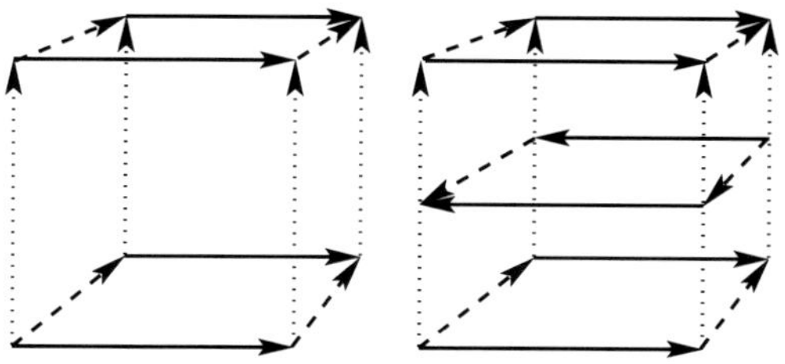

(a) G_1 (b) G_2

FIGURE 11. Polyhedra representing the groups G_1 and G_2.

G_4. $\Psi = \mathbb{Z}_4$ and Γ is generated by $\{\alpha, t_1, t_2, t_3\}$, where

$$\begin{cases} \alpha^4 = t_1 \\ \alpha t_2 \alpha^{-1} = t_3 \\ \alpha t_3 \alpha^{-1} = t_2^{-1} \end{cases} .$$

The generators $\{a_i\}$ are mutually orthogonal with $||a_2|| = ||a_3||$, while $\alpha = A \circ t_{a_1/4}$ with

$$\begin{cases} A(a_1) = a_1 \\ A(a_2) = a_3 \\ A(a_3) = -a_2 \end{cases} .$$

Here the situation is similar to the first two cases: we have a cube, we identify opposite vertical faces and glue top to bottom with a twist, this time of angle $\pi/2$. The Euler characteristic is zero and we get a manifold M_4, which is covered once by each block of Figure 12.

In these three cases the preimages of the manifolds, that is, the blocks in Figures 11 and 12, have quadrangles as the basis. Therefore, we can construct covering maps $g_1 : T^3 \to M_1$, $g_2 : T^3 \to M_2$ and $g_4 : T^3 \to M_4$ such that T^3 covers the manifold once, twice or four times, respectively.

Let us define mappings $F_i : \mathbb{R}^3 \to \mathbb{R}^3$, $i = 1, 2$ and 4, as $F_1 : x \mapsto 2x$, $F_2 : x \mapsto 3x$, and $F_4 : x \mapsto 5x$ for any $x \in \mathbb{R}^3$. Let $\pi_1 : \mathbb{R}^3 \to T^3$ be the usual covering map for the torus . The mappings F_i and π_1 induce a mapping F_i' on the torus. Next we show that F_i descends to a mapping $f_i : M_i \to M_i$. We can draw the following

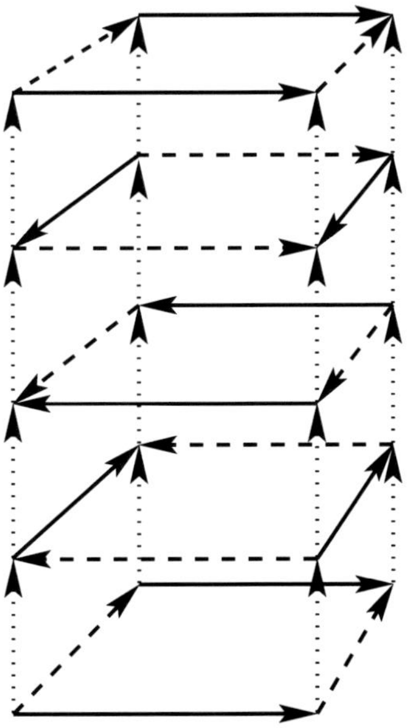

FIGURE 12. G_4

diagram, where $i = 1$, 2 or 4.

$$\begin{array}{ccc} \mathbb{R}^3 & \xrightarrow{F_i} & \mathbb{R}^3 \\ {\scriptstyle \pi_1}\downarrow & & \downarrow{\scriptstyle \pi_1} \\ T^3 & \xrightarrow{F_i'} & T^3 \\ {\scriptstyle g_i}\downarrow & & \downarrow{\scriptstyle g_i} \\ M_i & \xrightarrow{f_i} & M_i \end{array} \tag{25}$$

The mappings f_i, $i = 1,2$ or 4, are well defined and uniformly quasiregular of Lattès type. We prove these claims by using Theorem 3.9.

We first show that f_i is well defined. Let a corner point of the basis of a preimage of M_i on $\mathbb{R}^3$ be at the origin. Let m_i be a point on M_i. Denote one preimage point of the point m_i under the mapping $h_i := g_i \circ \pi_1$ with the help of the generators a_1, a_2 and a_3 as $x_1a_1 + x_2a_2 + x_3a_3$, where x_1, x_2 and $x_3 \in [0, 1)$.

Now the whole sets of preimage points are

$$h_1^{-1}(m_1) = \{(x_1 + n)a_1 + (x_2 + k)a_2 + (x_3 + l)a_3 \mid n, k, l \in \mathbb{Z}\},$$

$$\begin{aligned} h_2^{-1}(m_2) = \{&(x_1+2n)a_1+(x_2+k)a_2+(x_3+l)a_3, \\ &(x_1+1+2n')a_1+(1-x_2+k')a_2+(1-x_3+l')a_3 \\ &\mid n,n',k,k',l,l' \in \mathbb{Z}\}, \end{aligned}$$

$$\begin{aligned} h_4^{-1}(m_4) = \{&(x_1+4n)a_1+(x_2+k)a_2+(x_3+l)a_3, \\ &(x_1+1+4n')a_1+(1-x_2+k')a_2+(x_3+l')a_3 \\ &(x_1+2+4n'')a_1+(1-x_2+k'')a_2+(1-x_3+l'')a_3 \\ &(x_1+3+4n''')a_1+(x_2+k''')a_2+(1-x_3+l''')a_3 \\ &\mid n,n',n'',n''',k,k',k'',k''',l,l',l'',l''' \in \mathbb{Z}\}. \end{aligned}$$

(Note that the generators a_i in these sets vary depending on which manifold we are considering.) If we now apply the corresponding mappings F_i, we get the sets

$$F_1(h_1^{-1}(m_1)) = \{2(x_1+n)a_1+2(x_2+k)a_2+2(x_3+l)a_3 \mid n,k,l \in \mathbb{Z}\},$$

$$\begin{aligned} F_2(h_2^{-1}(m_2)) = \{&3(x_1+2n)a_1+3(x_2+k)a_2+3(x_3+l)a_3, \\ &3(x_1+1+2n')a_1+3(1-x_2+k')a_2+3(1-x_3+l')a_3 \\ &\mid n,n',k,k',l,l' \in \mathbb{Z}\}, \end{aligned}$$

$$\begin{aligned} F_4(h_4^{-1}(m_4)) = \{&5(x_1+4n)a_1+5(x_2+k)a_2+5(x_3+l)a_3, \\ &5(x_1+1+4n')a_1+5(1-x_2+k')a_2+5(x_3+l')a_3 \\ &5(x_1+2+4n'')a_1+5(1-x_2+k'')a_2+5(1-x_3+l'')a_3 \\ &5(x_1+3+4n''')a_1+5(x_2+k''')a_2+5(1-x_3+l''')a_3 \\ &\mid n,n',n'',n''',k,k',k'',k''',l,l',l'',l''' \in \mathbb{Z}\}. \end{aligned}$$

These sets belong to the sets $h_i^{-1}(p_i)$ for certain points p_i that have preimage points under h_i at $y_1a_1+y_2a_2+y_3a_3$, $y_1'a_1+y_2'a_2+y_3'a_3$ and $y_1'a_1+y_2'a_2+y_3'a_3$, respectively. Here $y_1 = 2(x_1+n)-k_1$ for some $k_1 \in \mathbb{Z}$ such that $y_1 \in [0,1)$, $y_2 = 2(x_2+k)-k_2$ for some $k_2 \in \mathbb{Z}$ such that $x_2 \in [0,1)$ and, $y_3 = 2(x_3+l)-k_3$ for some $k_3 \in \mathbb{Z}$ such that $x_3 \in [0,1)$. Similarly y_j' is the fractional part of the corresponding coefficient $3x_j$ in the set $F_2 \circ h_2^{-1}(m_2)$ and y_j'' is the fractional part of $5x_j$ in the set $F_4 \circ h_4^{-1}(m_4)$. So, each F_i descends to a mapping $f_i : M_i \to M_i$ that takes the points m_i to points p_i (for $i = 1$, 2 and 4). The mappings f_i are thus well defined.

In each of these three cases, $i = 1$, 2 or 4, the discrete groups Υ_i of isometries consist of all the translations between the preimages of the torus T^3 in $\mathbb{R}^3$ and the isometries between the preimages of M_i (respectively, one, two or four) inside a preimage of the torus. Thus the groups Υ_i are infinite, but the number of generators is finite in each case: three, four or nine, respectively.

Now the mapping $h_i = g_i \circ \pi_1$ is automorphic in the strong sense with respect to the group Υ_i, and for $F_i : x \mapsto (1+i)x$ it holds that

$$F_i \Upsilon_i F_i^{-1} \subset \Upsilon_i$$

for $i = 1$, 2 or 4. Therefore, by Theorem 3.9, the mappings f_i of $h_i(\mathbb{R}^3) = g_i \circ \pi_1(\mathbb{R}^3) = M_i$ are uniformly quasiregular as unique solutions to the respective Schröder equations.

If we take for the metric on M_i the induced flat metric from $\mathbb{R}^3$, the mapping f_i is, in addition, locally homeomorphic and conformal, since then there is no distortion in the mappings π_1 and g_i. The degrees of the mappings f_i are 8, 27 and 125 since each torus covers $(1+i)^3$ tori in the mapping F_i'.

G_6. $\Psi = \mathbb{Z}_2 \times \mathbb{Z}_2$ and Γ is generated by $\{\alpha, \beta, \gamma, t_1, t_2, t_3\}$, where $\gamma\beta\alpha = t_1 t_3$ and

$$\begin{array}{lll} \alpha^2 = t_1, & \alpha t_2 \alpha^{-1} = t_2^{-1}, & \alpha t_3 \alpha^{-1} = t_3^{-1}, \\ \beta t_1 \beta^{-1} = t_1^{-1}, & \beta^2 = t_2 & \beta t_3 \beta^{-1}, = t_3^{-1}, \\ \gamma t_1 \gamma^{-1} = t_1^{-1}, & \gamma t_2 \gamma^{-1} = t_2^{-1}, & \gamma^2 = t_3. \end{array}$$

The generators a_i are mutually orthogonal and $\alpha = A \circ t_{a_1/2}$ with

$$\begin{cases} A(a_1) = a_1 \\ A(a_2) = -a_2 \\ A(a_3) = -a_3 \end{cases},$$

$\beta = (B, t_{(a_2+a_3)/2})$ with

$$\begin{cases} B(a_1) = -a_1 \\ B(a_2) = a_2 \\ B(a_3) = -a_3 \end{cases},$$

and $\gamma = (B, t_{(a_1+a_2+a_3)/2})$ with

$$\begin{cases} C(a_1) = -a_1 \\ C(a_2) = -a_2 \\ C(a_3) = a_3 \end{cases}.$$

Again we have a cube and we identify its opposite faces, but now we glue top to bottom with a twist of angle π and also left face to right face with a twist of angle π. The schema gives a manifold M_6, since the Euler characteristic is $\chi(M_1) = 2 - 4 + 3 - 1 = 0$.

We can construct a uqr-mapping of Lattès type on M_6 in the same way as for M_1, M_2 and M_4. The only differences are that now the covering torus T^3 must contain the whole 2×2 block of Figure 13, and the discrete group Υ_6 of isometries has 6 different generators inside the preimage of the torus.

7.2.2. *Manifolds with hexagonal basis.* G_3. $\Psi = \mathbb{Z}_3$ and Γ is generated by $\{\alpha, t_1, t_2, t_3\}$, where

$$\begin{cases} \alpha^3 = t_1 \\ \alpha t_2 \alpha^{-1} = t_3 \\ \alpha t_3 \alpha^{-1} = t_2^{-1} t_3^{-1} \end{cases}.$$

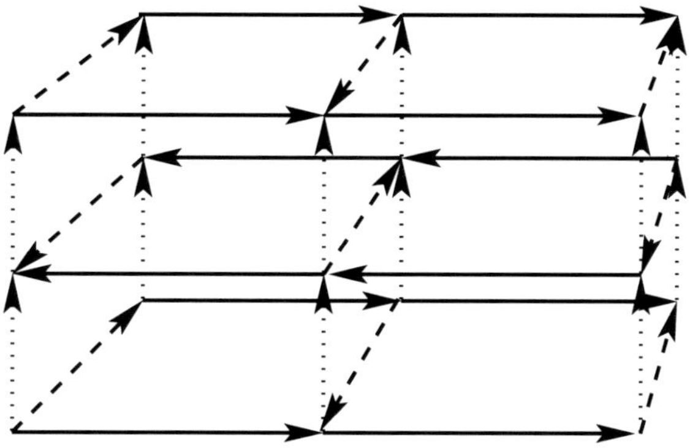

FIGURE 13. G_6

The generator a_1 is orthogonal to a_2 and a_3, $||a_2|| = ||a_3||$, $\{a_2, a_3\}$ is a hexagonal plane lattice (see Figure 10b), and $\alpha = A \circ t_{a_1/3}$ with

$$\begin{cases} A(a_1) = a_1 \\ A(a_2) = a_3 \\ A(a_3) = -a_2 - a_3 \end{cases}.$$

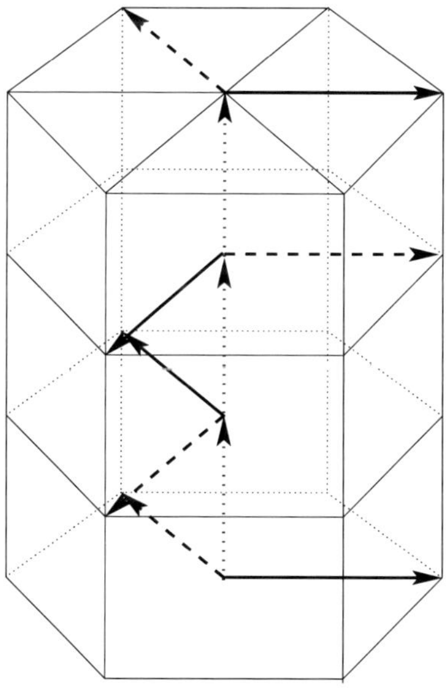

FIGURE 14. G_3

Consider one block of Figure 14. If we identify opposite vertical faces (according to the hexagonal plane lattice) and glue the top face to the bottom with a twist of angle $2\pi/3$, we get altogether two vertices, five edges, four faces and one cell for our polyhedron schema. Consequently, the Euler characteristic is $\chi = 2-5+4-1 = 0$, and we see that this schema gives us a manifold, say M_3.

G_5. $\Psi = \mathbb{Z}_6$ and Γ is generated by $\{\alpha, t_1, t_2, t_3\}$, where

$$\begin{cases} \alpha^6 = t_1 \\ \alpha t_2 \alpha^{-1} = t_3 \\ \alpha t_3 \alpha^{-1} = t_2^{-1} t_3 \end{cases} .$$

The generator a_1 is orthogonal to a_2 and a_3, $||a_2|| = ||a_3||$, $\{a_2, a_3\}$ is a hexagonal plane lattice, and $\alpha = A \circ t_{a_1/6}$ with

$$\begin{cases} A(a_1) = a_1 \\ A(a_2) = a_3 \\ A(a_3) = a_3 - a_2 \end{cases} .$$

Consider one block of Figure 15. Again we identify opposite vertical faces (according to the hexagonal plane lattice) and glue the top face to the bottom, now with a twist of angle $\pi/3$. We get altogether two vertices, five edges, four faces and one cell for our polyhedron schema. The Euler characteristic is $\chi = 2-5+4-1 = 0$, and we get a manifold M_5.

For M_3 and M_5 we can construct uqr-mappings of Lattès type as follows: We wish to find the smallest possible underlying fundamental domain of a torus lattice in $\mathbb{R}^3$. This domain is tiled with the 3 (or 6) stores high blocks of Figure 14 (or 15) according to the hexagonal plane lattice. The basis of the torus must cover at least one entire hexagon, the torus must be at least 3 (respectively 6) stores high, and we must be able to cover the space with similar tori. For example, the tori with basis as in Figure 16 fulfil our demands. These tori cover the manifold 3 (respectively 6) times.

Let us now construct Lattès type uqr-mappings on M_l, $l = 3$ or 5, with the help of these tori. Denote the covering map $g_l : T^3 \to M_l$. Let $F_3 : x \mapsto 4x$ and $F_5 : x \mapsto 7x$ for any $x \in \mathbb{R}^3$, and let $\pi_1 : \mathbb{R}^3 \to T^3$ be the usual covering map for the torus and F_3' and F_5' the maps they induce on the torus. As before, a mapping (let us denote it with f_l) is induced to the manifold. Now we have again the same diagram as in the tetragonal case:

$$\begin{array}{ccc} \mathbb{R}^3 & \xrightarrow{F_l} & \mathbb{R}^3 \\ \pi_1 \downarrow & & \downarrow \pi_1 \\ T^3 & \xrightarrow{F_l'} & T^3 \\ g_l \downarrow & & \downarrow g_l \\ M_l & \xrightarrow{f_l} & M_l \end{array} \tag{26}$$

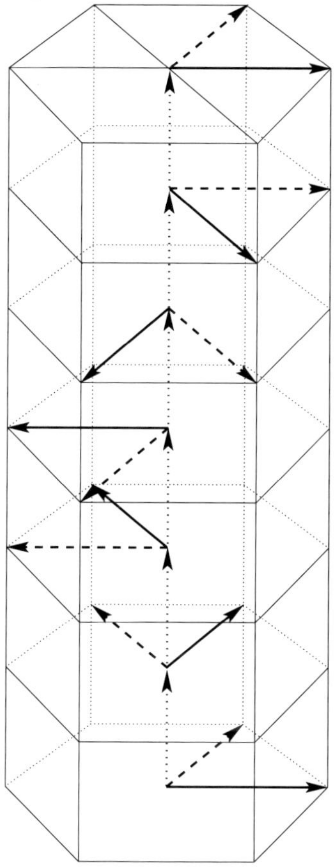

FIGURE 15. G_5

The mapping $f_l : M_l \to M_l$ is well defined and uniformly quasiregular of Lattès type — in fact a locally homeomorphic conformal mapping can be made, which is far more than we need. We prove these claims by using Theorem 3.9.

We first show that f_l is well defined. Let the middle point of the basis of a preimage of M_l on $\mathbb{R}^3$ be at the origin. Let m_l be a point on M_l. Denote one preimage point of the point m_l under the mapping $h_l = g_l \circ \pi_1$ with the help of the generators a_1, a_2 and a_3 as $x_1a_1 + x_2a_2 + x_3a_3$, where $x_1 \in [0, 1)$, $x_2 \in [-1, 1)$ and $x_3 \in [-1, 1)$. The preimages of these points m_l on $\mathbb{R}^3$ under the mappings h_l are sets of points

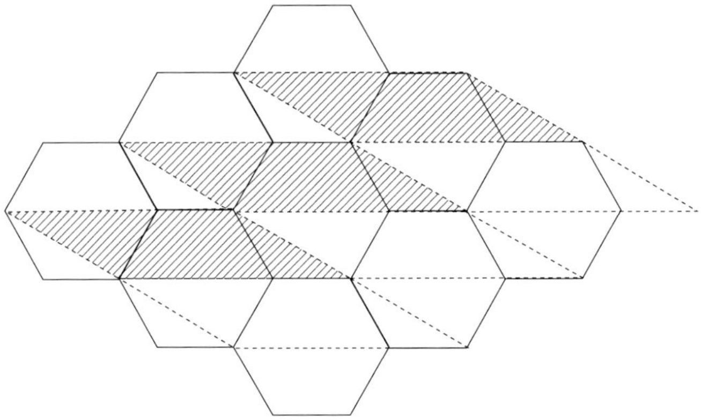

FIGURE 16. Basis for the tori in hexagonal lattice.

$$\begin{aligned} h_3^{-1}(m_3) = \{&(x_1+3n)a_1+(x_2+k)a_2+(x_3+s)a_3,\\ &(x_1+1+3n')a_1+(-x_3+k')a_2+(x_2-x_3+s')a_3\\ &(x_1+2+3n'')a_1+(x_3-x_2+k'')a_2+(-x_2+s'')a_3\\ &\mid n,n',n'',k,k',k'',s,s',s''\in\mathbb{Z}\}, \end{aligned}$$

$$\begin{aligned} h_5^{-1}(m_5) = \{&(x_1+6n)a_1+(x_2+k)a_2+(x_3+s)a_3,\\ &(x_1+1+6n')a_1+(-x_3+k')a_2+(x_2+x_3+s')a_3\\ &(x_1+2+6n'')a_1+(-x_2-x_3+k'')a_2+(x_2+s'')a_3\\ &(x_1+3+6n''')a_1+(-x_2+k''')a_2+(-x_3+s''')a_3\\ &(x_1+4+6\tilde{n})a_1+(x_3+\tilde{k})a_2+(-x_2-x_3+\tilde{s})a_3\\ &(x_1+5+6\bar{n})a_1+(x_2+x_3+\bar{k})a_2+(-x_2+\bar{s})a_3\\ &\mid n,n',n'',n''',\tilde{n},\bar{n},k,k',k'',k''',\tilde{k},\bar{k},s,s',s'',s''',\tilde{s},\bar{s}\in\mathbb{Z}\}. \end{aligned}$$

If we now apply the mappings F_l, we again get sets, which are mapped onto one point by h_l: The whole set $F_3(h_3^{-1}(m_3))$ is mapped by h_3 to a point p_3, which has $y_1a_1+y_2a_2+y_3a_3$, where y_j is the fractional part of the corresponding coefficient $4x_j$ in the set $F_3(h_3^{-1}(m_3))$, as one preimage. Similarly the set $F_5(h_5^{-1}(m_5))$ is mapped by h_5 to a point p_5, which has $y_1'a_1+y_2'a_2+y_3'a_3$, where y_j is the fractional part of the corresponding coefficient $7x_j$ in the set $F_5(h_5^{-1}(m_5))$, as one preimage. Again we see that both F_l descend to a mapping $f_l : M_l \longrightarrow M_l$ and the mappings f_l, $l=3$ or 5, are well defined.

In the case of M_5, Υ_5, the discrete group of isometries, consists of all the translations between the preimages of the torus T^3 in $\mathbb{R}^3$ and the isometries between the six preimages of M_5 inside a preimage of the torus. Thus the group Υ_5 is infinite, but the number of its generators is finite: three generators for the translations and only a finite number for the isometries inside a preimage of the torus. For M_3, the discrete group of isometries, Υ_3, is even more simple: it has the same three generators outside and only three inside the preimage of the torus.

Now $h_l = g_l \circ \pi_1$ is automorphic in the strong sense with respect to the group Υ_l, and for F_l it holds that

$$F_l \Upsilon_l F_l^{-1} \subset \Upsilon_l.$$

Therefore, by Theorem 3.9, the mappings f_l of $h_l(\mathbb{R}^3) = g_l \circ \pi_1(\mathbb{R}^3) = M_l$ are uniformly quasiregular. If we take for the metric on M_l the induced flat metric from $\mathbb{R}^3$, the mappings f_l are, in addition, locally homeomorphic and conformal, since then there is no distortion in the mappings π_1 and g_l. The degrees of the mappings f_3 and f_5 are $4^3 = 64$ and $7^3 = 343$.

The Julia sets for all these mappings related to euclidean space forms are chaotic, that is, their Fatou set is empty, since there are no completely invariant sets for the uniformly quasiregular mappings which we have constructed. We prove this for the case G_1, the other ones follow analogously. See [Ma1] for similar considerations in $\overline{\mathbb{R}}^n$.

Lemma 7.5. *The uniformly quasiregular mapping $f_1 : M_1 \to M_1$ is chaotic.*

Proof. The origin is a repelling fixed point for the mapping $F : x \mapsto 2x$. Consider its Υ-orbit, $\Upsilon(0) = \{\gamma(0) \mid \gamma \in \Upsilon\} = 2\mathbb{Z}^3$. The set $E = \cup_{k\geq 0} F^{-k}(\Upsilon(0))$ is a dense subset of $\mathbb{R}^3$. Hence, $h_1(E)$ is a dense subset of M_1. We conclude that (f_1^k) cannot be equicontinuous in a neighbourhood of any point of M_1. This means that the Fatou set is empty and the mapping f_1 is chaotic. □

7.3. **Manifolds covered by $\mathbb{S}^2 \times \mathbb{R}$.** There are exactly four compact 3-manifolds which have $\mathbb{S}^2 \times \mathbb{R}$ as the riemannian covering space. Two of them, namely, the twisted bundle $\mathbb{S}^2 \tilde{\times} \mathbb{S}$ and the projective plane bundle $\mathbb{P}^2 \times \mathbb{S}$, are non-orientable. Therefore we need to consider only two manifolds, the sphere bundle $\mathbb{S}^2 \times \mathbb{S}$ and the connected sum of two projective 3-spaces $\mathbb{P}^3 \# \mathbb{P}^3$. [Jo, p. 18]

7.3.1. *Sphere bundle $\mathbb{S}^2 \times \mathbb{S}$.* In this chapter we will construct a branched uqr-mapping to $\mathbb{S}^2 \times \mathbb{S}$. Consider $\mathbb{S}^2 \times \mathbb{S}$ as a sphere $\mathbb{S}^2$, which has a line segment attached to each of its points and the ends of each line segment identified (see Figure 17(a)). Now we see that $\mathbb{S}^2 \times \mathbb{S}$ has $\mathbb{R}^3$ as a branched cover: let us divide $\mathbb{R}^3$ into cubes. The base of every other cube is identified with the upper half space of $\mathbb{S}^2$, every other with the lower half space (denoted by + and - in Figures 17(a) and 17(b)). The top of the cube is identified with the bottom, thus forming $\mathbb{S}$ to each point of $\mathbb{S}^2$.

To be able to tile $\mathbb{R}^3$ with $\mathbb{S}^2 \times \mathbb{S}$, we have to make one reproducible block from two images of $\mathbb{S}^2 \times \mathbb{S}$ as in Figure 17(b). One building block, which thereby covers $\mathbb{S}^2 \times \mathbb{S}$ twice, is presented in Figure 18.

Let us now define which points in $\mathbb{R}^3$ need to be identified to present the whole $\mathbb{R}^3$ as a branched cover for $\mathbb{S}^2 \times \mathbb{S}$. Let e_1, e_2 and e_3 be the three orthogonal unit

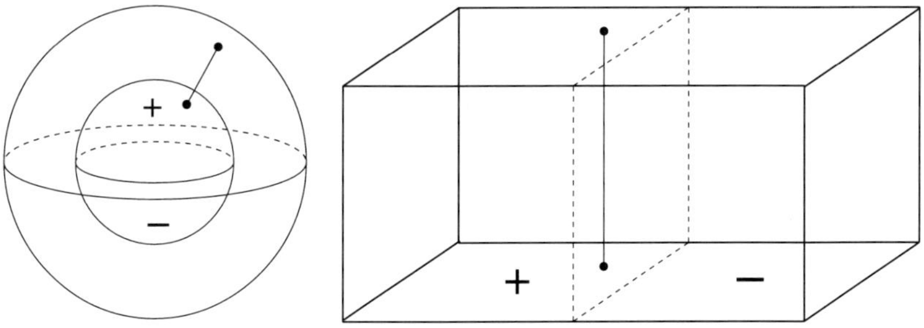

(a) $\mathbb{S}^2 \times \mathbb{S}$ (b) One block in $\mathbb{R}^3$ that covers $\mathbb{S}^2 \times \mathbb{S}$.

FIGURE 17. How to see $\mathbb{S}^2 \times \mathbb{S}$ as a block in $\mathbb{R}^3$.

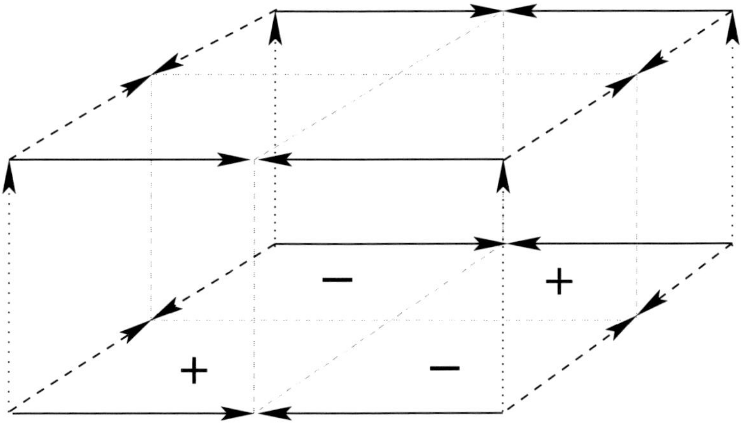

FIGURE 18. One block which covers $\mathbb{S}^2 \times \mathbb{S}$ twice.

vectors in $\mathbb{R}^3$, and let x_1, x_2 and $x_3 \in \mathbb{R}$ be an arbitrary point in $\mathbb{R}^3$. We identify all the points of the set

$$\begin{aligned}&\{(x_1+n)e_1+(x_2+2m)e_2+(x_3+2k)e_3,\\ &(x_1+n')e_1+(2-x_2+2m')e_2+(2-x_3+2k')e_3\\ &\mid n,n',m,m',k,k'\in\mathbb{Z}\}\end{aligned} \tag{27}$$

to be the same point of the manifold. This identification presents $\mathbb{R}^3$ as a branched cover for $\mathbb{S}^2 \times \mathbb{S}$, see Figure 18. The branch set is the skeleton of the tiling.

The situation is now fundamentally different from the previous ones, since we have branching on each vertical line of Figure 18. We can, however, proceed analogically with the previous cases. Define the covering map to be $g : \mathbb{R}^3 \to \mathbb{S}^2 \times \mathbb{S}$. We use again the mappings $F : \mathbb{R}^3 \to \mathbb{R}^3$, where $F : x \mapsto 2x$ for any $x \in \mathbb{R}^3$. The mappings

F and g induce a mapping f to the manifold. We can draw the following diagram:

$$\begin{array}{ccc} \mathbb{R}^3 & \xrightarrow{F} & \mathbb{R}^3 \\ {\scriptstyle g}\downarrow & & \downarrow{\scriptstyle g} \\ \mathbb{S}^2\times\mathbb{S} & \xrightarrow{f} & \mathbb{S}^2\times\mathbb{S} \end{array} \tag{28}$$

The mapping f is a well-defined and uniformly quasiregular mapping of Lattès type, which we can prove in the same manner as in the previous cases. Let $\{e_1, e_2, e_3\}$ be an orthonormal basis for $\mathbb{R}^3$ and fix the origin to be at the bottom-left corner of Figure 18. Let x be such a point on $\mathbb{S}^2 \times \mathbb{S}$ that $x_1e_1 + x_2e_2 + x_3e_3$, where $x_1 \in [0, 1]$ and x_2, $x_3 \in [0, 2)$, is one of its preimages under g. Now we see that the whole preimage set of the point x under the covering map g is

$$\begin{aligned} g^{-1}(x) &= \{(x_1+n)e_1 + (x_2+2m)e_2 + (x_3+2k)a_3 \mid n,m,k \in \mathbb{Z}\} \\ &\quad \cup \{(x_1+n')e_1 + (2-x_2+2m')e_2 + (2-x_3+2k')e_3 \\ &\qquad \mid n',m',k' \in \mathbb{Z}\}. \end{aligned}$$

Consequently,

$$\begin{aligned} F(g^{-1}(x)) &= \{2(x_1+n)e_1 + 2(x_2+2m)e_2 + 2(x_3+2k)a_3 \mid n,m,k \in \mathbb{Z}\} \\ &\quad \cup \{2(x_1+n')e_1 + 2(2-x_2+2m')e_2 + 2(2-x_3+2k')e_3 \\ &\qquad \mid n',m',k' \in \mathbb{Z}\} \\ &\subset \{(2x_1+n)e_1 + (2x_2+2m)e_2 + (2x_3+2k)a_3 \mid n,m,k \in \mathbb{Z}\} \\ &\quad \cup \{(2x_1+n')e_1 + (2-2x_2+2m')e_2 + (2-2x_3+2k')e_3 \\ &\qquad \mid n',m',k' \in \mathbb{Z},\} \end{aligned}$$

which under the mapping g is again just one point $y := y_1e_1 + y_2e_2 + y_3e_3$, where y_i is the fractional part of $2x_i$, $i = 1, 2, 3$. Thus F descends to a well defined mapping $f : \mathbb{S}^2 \times \mathbb{S} \to \mathbb{S}^2 \times \mathbb{S}$.

In this case, the discrete group of isometries, Υ, consists of all the translations between the four-parted blocks (see Figure 18) in $\mathbb{R}^3$ and the 180-degree rotation around the middle axis of Figure 18. Thus the group Υ has four generators. The mapping g is automorphic in the strong sense with respect to the group Υ, and for $F : x \mapsto 2x$ it holds that

$$F\Upsilon F^{-1} \subset \Upsilon.$$

Therefore, by Theorem 3.9, the well-defined mapping $f : \mathbb{S}^2\times\mathbb{S} \to \mathbb{S}^2\times\mathbb{S}$ is uniformly quasiregular. The degree of the mapping f is 8.

Note that for this manifold we can also construct such a uniformly quasiregular mapping which has no branching: Consider $\mathbb{S}^2 \times \mathbb{R}$ as a cover for $\mathbb{S}^2 \times \mathbb{S}$ in such a way that $\mathbb{S}^2 \times \mathbb{R}$ is divided to annuli, each of which covers $\mathbb{S}^2 \times \mathbb{S}$ once by a covering map $\pi : \mathbb{S}^2 \times \mathbb{R} \to \mathbb{S}^2 \times \mathbb{S}$, $(z, t) \mapsto (z, e^{i\pi t})$. Let $F : \mathbb{S}^2 \times \mathbb{R} \to \mathbb{S}^2 \times \mathbb{R}$ take (z, t) to $(z, 2t)$. Then the equation $f \circ \pi = \pi \circ F$ defines a mapping f, which is uniformly quasiregular and locally - but not globally - homeomorphic.

7.3.2. *Connected sum* $\mathbb{P}^3\#\mathbb{P}^3$. We will again construct a branched uqr-mapping. We recall that a projective space is the space of one-dimensional vector subspaces of a given vector space. The notation $\mathbb{P}^n$ denotes the real projective space of dimension n (i.e., the space of one-dimensional vector subspaces of $\mathbb{R}^{(n+1)}$). $\mathbb{P}^n$ can also be viewed as the set $\mathbb{S}^n/\{x,-x\}$, where x and $-x$ denote the antipodal points.

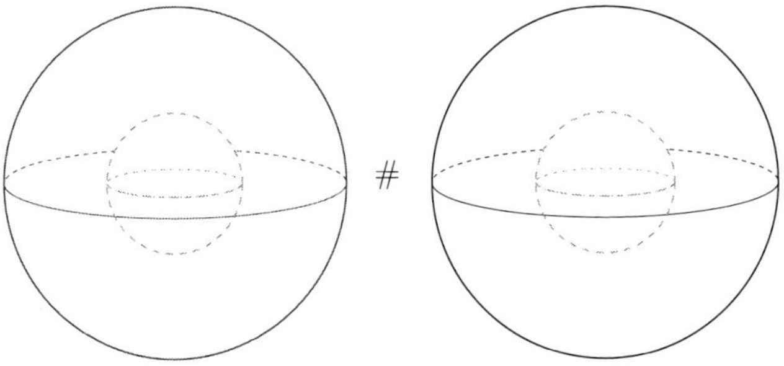

FIGURE 19. $\mathbb{P}^3\#\mathbb{P}^3$

The manifold $\mathbb{P}^3\#\mathbb{P}^3$ is obtained by identifying diametrical points of the boundary spheres K_1 and K_2 of $\mathbb{S}^2 \times I$ (see Figures 19 and 20) [ST, p. 417]. The dotted 2-sphere separates this manifold into two punctured projective spaces. The fibres are the radii of $\mathbb{S}^2 \times I$; any two diametrical radii form one fibre.

Let us look at the projective space $\mathbb{P}^3\#\mathbb{P}^3$ as a block in $\mathbb{R}^3$, similarly as we did for $\mathbb{S}^2 \times \mathbb{S}$ in Figures 17(a) and 17(b). Now we have to be even more careful to get the identifications at the boundaries right. Using the same symbols and colours as in Figure 20, we obtain Figure 21. Thus in Figure 21 any two adjacent cubes cover the space $\mathbb{P}^3\#\mathbb{P}^3$ once. From Figure 21 we can count the Euler characteristic to be $\chi = 2-6+8-4 = 0$, as it should be.

Now we can define which points on $\mathbb{R}^3$ need to be identified to present $\mathbb{R}^3$ as a branched cover for $\mathbb{P}^3\#\mathbb{P}^3$ as follows: Let e_1, e_2 and e_3 be the three orthogonal unit vectors in $\mathbb{R}^3$ and let $x_1e_1+x_2e_2+x_3e_3$ be an arbitrary point in $\mathbb{R}^3$, x_1, x_2, $x_3 \in \mathbb{R}$. We need to identify all the points in the set

$$
\begin{aligned}
&\{(x_1+2n)e_1+(x_2+2m)e_2+(x_3+2k)e_3,\\
&(x_1+2n')e_1+(2-x_2+2m')e_2+(2-x_3+2k')e_3\\
&(2-x_1+2n'')e_1+(1+x_2+2m'')e_2+(1-x_3+2k'')e_3 \qquad (29)\\
&(2-x_1+2n''')e_1+(1-x_2+2m''')e_2+(1+x_3+2k''')e_3\\
&\mid n, n'n,'', n''', m, m', m'', m''', k, k', k', k''' \in \mathbb{Z}\}
\end{aligned}
$$

to be the same point on $\mathbb{P}^3\#\mathbb{P}^3$.

We have again branching on each vertical line of Figure 21. Denote the covering map defined above to be $g : \mathbb{R}^3 \to \mathbb{P}^3\#\mathbb{P}^3$. We again use the mapping $F : \mathbb{R}^3 \to \mathbb{R}^3$,

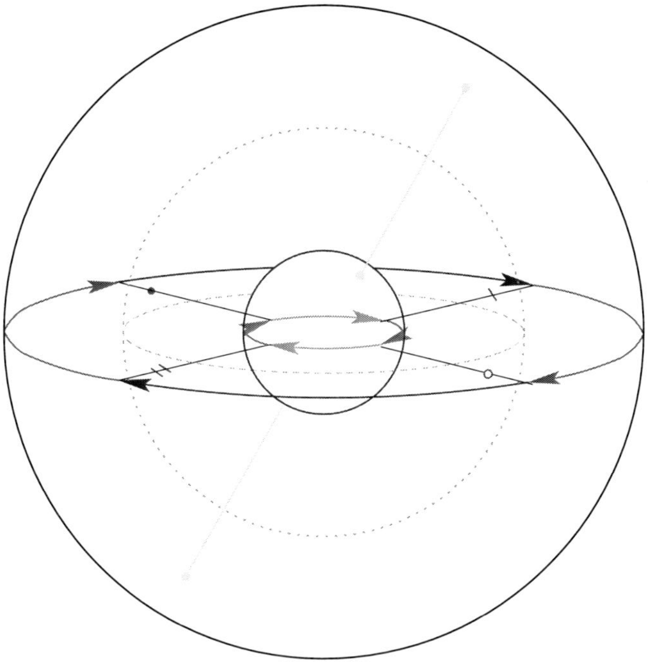

FIGURE 20. One fibre of $\mathbb{P}^3\#\mathbb{P}^3$ emphasized in yellow.

where $F : x \mapsto 2x$ for any $x \in \mathbb{R}^3$. The mappings F and g again induce a mapping f to the manifold:

$$\begin{array}{ccc} \mathbb{R}^3 & \xrightarrow{F} & \mathbb{R}^3 \\ g\downarrow & & \downarrow g \\ \mathbb{P}^3\#\mathbb{P}^3 & \xrightarrow{f} & \mathbb{P}^3\#\mathbb{P}^3 \end{array} \tag{30}$$

Let us check again that the mapping f is a well-defined and uniformly quasiregular mapping of Lattès type. Fix the origin to be at the bottom-left corner of Figure 21 and denote the orthogonal unit vectors again by e_1, e_2 and e_3. Let x be such a point on $\mathbb{P}^3\#\mathbb{P}^3$ that $x_1e_1 + x_2e_2 + x_3e_3$, where $x_1 \in [0, 2)$, x_2, $x_3 \in [0, 1)$, is one of

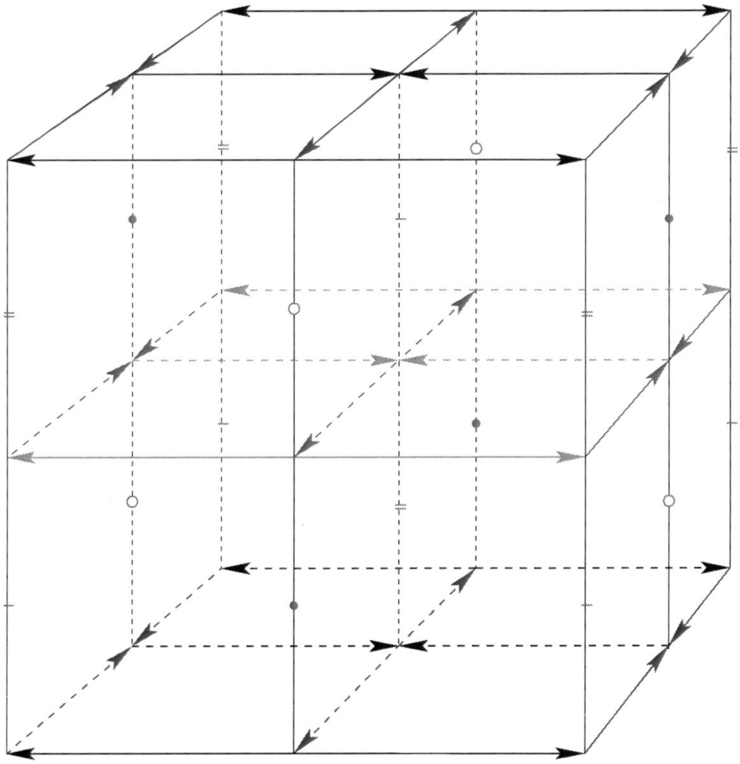

FIGURE 21. Representation of $\mathbb{P}^3\#\mathbb{P}^3$ by polyhedra.

its preimages under g. Then the set of preimage points of x is

$$\begin{aligned} g^{-1}(x) &= \{(x_1+2n)e_1+(x_2+2m)e_2+(x_3+2k)a_3 \mid n,m,k\in\mathbb{Z}\} \\ &\cup \{(x_1+2n')e_1+(2-x_2+2m')e_2+(2-x_3+2k')e_3 \\ &\quad \mid n',m',k'\in\mathbb{Z}\} \\ &\cup \{(2-x_1+2n'')e_1+(1+x_2+2m'')e_2+(1-x_3+2k'')e_3 \\ &\quad \mid n'',m'',k''\in\mathbb{Z}\} \\ &\cup \{(2-x_1+2n''')e_1+(1-x_2+2m''')e_2+(1+x_3+2k''')e_3 \\ &\quad \mid n''',m''',k'''\in\mathbb{Z}\}. \end{aligned}$$

Consequently,

$$\begin{aligned}
&F(g^{-1}(x)) = \\
&\{2(x_1 + 2n)e_1 + 2(x_2 + 2m)e_2 + 2(x_3 + 2k)a_3 \mid n, m, k \in \mathbb{Z}\} \\
&\cup \{2(x_1 + 2n')e_1 + 2(2 - x_2 + 2m')e_2 + 2(2 - x_3 + 2k')e_3 \\
&\quad \mid n', m', k' \in \mathbb{Z}\} \\
&\cup \{2(2 - x_1 + 2n'')e_1 + 2(1 + x_2 + 2m'')e_2 + 2(1 - x_3 + 2k'')e_3 \\
&\quad \mid n'', m'', k'' \in \mathbb{Z}\} \\
&\cup \{2(2 - x_1 + 2n''')e_1 + 2(1 - x_2 + 2m''')e_2 + 2(1 + x_3 + 2k''')e_3 \\
&\quad \mid n''', m''', k''' \in \mathbb{Z}\},
\end{aligned}$$

which under the mapping g is again just one point $y := y_1e_1 + y_2e_2 + y_3e_3$, where y_i is the fractional part of $2x_i$, $i = 1, 2, 3$. Thus F descends to a well defined mapping $f : \mathbb{P}^3\#\mathbb{P}^3 \to \mathbb{P}^3\#\mathbb{P}^3$.

The discrete group of isometries, Υ, now consists of all the translations between the eight-parted blocks (see Figure 21) in $\mathbb{R}^3$ and the 180-degree rotation around the vertical middle axis of Figure 21 and of the reflection in terms of the vertical middle axis of one of the eight blocks. Thus the group Υ has nine generators. The mapping g is automorphic in the strong sense with respect to the group Υ, and for $F : x \mapsto 2x$ it holds that

$$F\Upsilon F^{-1} \subset \Upsilon.$$

Therefore, by Theorem 3.9, the mapping $f : \mathbb{P}^3\#\mathbb{P}^3 \to \mathbb{P}^3\#\mathbb{P}^3$ is uniformly quasiregular.

Julia sets for these uniformally quasiconformal mappings which we constructed for $\mathbb{S}^2 \times \mathbb{S}$ and $\mathbb{P}^3\#\mathbb{P}^3$ are chaotic. This can be proved in the same way as Lemma 7.5. Note that in the same manner as for the manifold $\mathbb{S}^2 \times \mathbb{R}$, we could also for this manifold construct a uniformly quasiregular mapping with an empty branch set.

Remark 7.6. *We have now constructed uniformly quasiregular mappings on all orientable quasiregularly elliptic 3-dimensional compact riemannian manifolds, that is, on manifolds covered by $\mathbb{S}^3$, $\mathbb{S}^2 \times \mathbb{R}$ or $\mathbb{R}^3$. Thus we have proved Theorem 7.1.*

References

[An] ANDERSON, M. T.: *Scalar curvature and geometrization conjectures for 3-manifolds.* M.S.R.I. Publ. vol.30 (1997), 49-82.

[Bi] BISHOP, R. L. AND R. J. CRITTENDEN: *Geometry of Manifolds.* AMS Chelsea Publishing, 2001.

[BH] BONK, M. AND J. HEINONEN: *Quasiregular mappings and cohomology.* Acta Math. 186 (2001), 219-238.

[Ch1] CHERNAVSKIĬ, A. V.: *Finite-to-one open mappings of manifolds.* (Russian). Mat. Sb. 65 (1964) 357-369.

[Ch2] CHERNAVSKIĬ, A. V.: *Remarks on the paper "On finite-to-one mappings of manifolds".* (Russian). Mat. Sb. 66 (1965) 471-472.

[CT] CHURCH, P. T. AND J. G. TIMOURIAN: *Differentiable maps with small critical set or critical set image.* Indiana Univ. Math. J. 27 (1978), 953-971.

[Ge] GEHRING, F. W.: *Rings and quasiconformal mappings in space.* Trans. Amer. Math. Soc. 103 (1962) 353-393.

[Gr] GROMOV, M.: *Structures métriques pour les variétés riemanniennes.* Edited by J. Lafontaine and P. Pansu, CEDIC, Paris, 1981.

[Hi] HICKS, N. J.: *Notes on Differential Geometry.* D. Van Nostrand Company, Inc., 1965.

[HMM] HINKKANEN, A., G. J. MARTIN AND V. MAYER: *Local dynamics of uniformly quasiregular mappings.* Math. Scand. 95 (2004), nro. 1, 80-100.

[IM1] IWANIEC, T. AND G. J. MARTIN: *Geometric Function Theory and Non-linear Analysis.* Oxford mathematical monographs, Clarendon Press, 2001.

[IM2] IWANIEC, T. AND G. J. MARTIN: *Quasiregular semigroups.* Ann. Acad. Sci. Fenn. Math. Vol. 21 (1996), 241-254.

[JS] JACO, W. AND P. SHALEN: *Seifert fibered spaces in 3-manifolds.* Mem. Amer. Math. Soc. 21 (1979), no. 220, 192 pp.

[Jo] JOHANNSON, K.: *Homotopy equivalence of 3-manifolds with boundaries.* Lecture Notes in Math, 761, Springer, Berlin, (1979).

[Kn] KNESER, H.: *Geschlossene Flächen in dreidimensionalen Mannigfaltigkeiten.* Jahres. Deut. Math. Verein. 38 (1929), 248-260.

[Jo] JORMAKKA, J.: *The existence of quasiregular mappings from R^3 to closed orientable 3-manifolds.* Ann. Acad. Sci. Fenn. Ser. A I Math. Dissertationes No. 69 (1988).

[Ka] KANGASLAMPI, R.: *Tasaisesti kvasisäännölliset kuvaukset ja elliptiset Riemannin monistot.* Licentiate's thesis. Helsinki University of Technology, 2006.

[Lee] LEE, J. M.: *Riemannian Manifolds.* An Introduction to Curvature. Springer-Verlag, 1997.

[LV] LEHTO, O. AND K. I. VIRTANEN: *Quasiconformal Mappings in the Plane.* Springer-Verlag, 1973.

[MMP] MARTIN, G. J., V. MAYER AND K. PELTONEN: *The generalized Lichnerowicz problem: uniformly quasiregular mappings and space forms.* Proc. Amer. Math. Soc. 134 (2006), no. 7, 2091-2097.

[MRV1] MARTIO, O., S. RICKMAN AND J. VÄISÄLÄ: *Definitions for quasiregular mappings.* Ann. Acad. Sci. Fenn. Ser. AI 448 (1969), 1-40.

[MRV2] MARTIO, O., S. RICKMAN AND J. VÄISÄLÄ: *Distortion and singularities of quasiregular mappings.* Ann. Acad. Sci. Fenn. Ser. AI 465 (1970), 1-13.

[MRV3] MARTIO, O., S. RICKMAN AND J. VÄISÄLÄ: *Topological and metric properties of quasiregular mappings.* Ann. Acad. Sci. Fenn. Ser. AI 488 (1971), 1-31.

[Ma1] MAYER, V.: *Uniformly quasiregular mappings of Lattès type.* Conform. Geom. Dyn. 1 (1997), 104-111.

[Ma2] MAYER, V.: *Phénomènes de rigidité en dynamique holomorphe et quasirégulière, ensembles Lip-homogènes.* Habilitation à Diriger des Recherches en Sciences Mathématiques. Université des Sciences et Technologies de Lille, 2000.

[Mil] MILNOR, J.: *A unique factorization theorem for 3-manifolds.* Amer. J. Math. 84 (1962), 1-7.

[Min] MINIOWITZ, R.: *Normal families of quasimeromorphic mappings.* Proc. Amer. Math. Soc. Vol. 84/1 (1982), 35-43.

[Mo] MONTESINOS, J. M.: *Classical tessellations and three-manifolds.* Springer-Verlag, 1987.

[Pe] PELTONEN, K.: *Examples of uniformly quasiconformal mappings.* Conformal Geometry and Dynamics Vol. 3 (1999), 158-163.

[Re] RESHETNYAK, YU. G.: *The Liouville theorem with minimial regularity conditions.* (Russian). Sibirsk. Mat. Zh. 8 (1967) 835-840.

[R1] RICKMAN, S.: *The analogue of Picard's theorem for quasiregular mappings in dimension three.* Acta Math. 154 (1985), 195-242.

[R2] RICKMAN, S.: *Quasiregular Mappings.* Ergebnisse der Mathematik 26. Springer-Verlag, 1993.

[R3] RICKMAN, S.: *Simply connected quasiregularly elliptic 4-manifolds.* Ann. Acad. Sci. Fenn. Math. 31 (2006), no. 1, 97-110.

[ST] SEIFERT, H. AND W. THRELFALL: *A textbook of topology.* Including Seifert: Topology of 3-dimensional fibered spaces. Academic Press Inc., New York, 1980.

[Sch] SCHIFF, J.: *Normal Families.* Universitext. Springer-Verlag, 1993.

[Sco] SCOTT, P.: *The geometries of 3-manifolds.* Bull. London. Math. Soc. 15 (1983), 401-487.

[St] STILLWELL, J.: *Classical topology and combinatorial group theory.* Springer-Verlag, 1980.

[Th] THURSTON, W. P.: *Three-Dimensional Geometry and Topology.* Princeton University Press, 1997.

[Tu] TUCKER, A.: *Branched and folded coverings.* Bull. Amer. Math. Soc. 42 (1936), 859-862.

[VSC] VAROPOULOS, N., L. SALOFF-COSTE AND T. COULHON: *Analysis and Geometry on Groups.* Cambridge University Press, 1992.

[V1] VÄISÄLÄ, J.: *Discrete open mappings on manifolds.* Ann. Acad. Sci. Fenn. Ser. A I Math. 392 (1966), 1-10.

[V2] VÄISÄLÄ, J.: *Topologia II.* Limes ry, 1999.

[Wo] WOLF, J. A.: *Spaces of constant curvature.* 5th edition. Publish or Perish, Inc. 1984.

[Za] ZALCMAN, L.: *A heuristic principle in complex function theory.* Amer. Math. Monthly 82 (1975), 813-817.

DEPARTMENT OF MATHEMATICS AND SYSTEMS ANALYSIS, P.O. BOX 1100 (OTAKAARI 1M), FI-02015 HELSINKI UNIVERSITY OF TECHNOLOGY, FINLAND

E-mail address: riikka.kangaslampi@tkk.fi